Composites Engineering: An A–Z Guide

Composites Engineering: An A–Z Guide

Parvez Alam
School of Engineering, Institute for Materials and Processes, The University of Edinburgh, Edinburgh, UK

IOP Publishing, Bristol, UK

ISBN 978-0-7503-3527-0 (ebook)
ISBN 978-0-7503-3525-6 (print)
ISBN 978-0-7503-3528-7 (myPrint)
ISBN 978-0-7503-3526-3 (mobi)

DOI 10.1088/978-0-7503-3527-0

Version: 20211001

IOP ebooks

British Library Cataloguing-in-Publication Data: A catalogue record for this book is available from the British Library.

Published by IOP Publishing, wholly owned by The Institute of Physics, London

IOP Publishing, Temple Circus, Temple Way, Bristol, BS1 6HG, UK

US Office: IOP Publishing, Inc., 190 North Independence Mall West, Suite 601, Philadelphia, PA 19106, USA

Monpura

Contents

Preface

With composite materials experiencing a continued rise in interest, application and utility, there has never been a better time to prepare an informative A–Z guide for inductees to the world of composites engineering. While the purpose of this book is essentially to help newcomers to composites engineering, there are also benefits anticipated for intermediate to more established composites engineers seeking quick reference in areas outside of their normal expertise or practice. Additionally, there is an increasing number of university courses requiring foundational knowledge in the area of composites science, technology, manufacturing and engineering. As such, there is clear justification for the production of textual resources that can provide quick information on different aspects and terminologies unique to, or important in composites engineering. Since there are no such resources available at present, this book aims to fill the literary gap. The book is aimed at students and industry-based engineers alike, and the premise of the book is to provide more information than would normally be obtained from a dictionary or a glossary of terms, whilst avoiding lengthy explanations normally found in thorough and descriptive tomes.

Author biography

Parvez Alam

Parvez Alam is an academic at The University of Edinburgh, UK. He is a Fellow of the Institute of Materials, Minerals and Mining, a Fellow of the Institution of Mechanical Engineers, a Fellow of the Royal Society of Biology, and a Chartered Mechanical Engineer. He has researched a wide variety of both engineering and biological composites since the late 1990s, currently focussing on topics related to biomimetic design, mechanical metamaterials, comparative biomechanics, machine design, and composites engineering. At Edinburgh, he lectures courses on engineering design, having previously developed courses in composites engineering, computational modelling and biomimetics while in the Nordic and South East Asian regions.

IOP Publishing

Composites Engineering: An A–Z Guide

Parvez Alam

Chapter 1

'A'

1.1 Abrasion

Composites may be subject to abrasive wear (abrasion), where material loss is a result of the passage of a solid surface on a solid surface (sliding/rubbing/friction). It should be noted that this is fundamentally different to erosive wear (erosion), where material loss is a function of solids impacting a solid surface. Abrasion is a phenomenon affecting the entire range of composites including polymer matrix composites (PMCs), metal matrix composites (MMCs) and ceramic matrix composites (CMCs). Examples where composites may experience abrasion include but are not limited to: mastication of dental composites (PMCs or CMCs) [1], interactions between mechanical components (brake rotors, rod and piston, gearing, pulleys, bearings) (usually MMCs or metal alloys) [2], machining processes (CMCs) [3], and bulk solid transportation [4] (e.g. mining tools).

Since abrasion is a solid-to-solid contact phenomenon, the properties of material(s) responsible for abrading composites are as important as the properties of the composite itself. An understanding of the abrasive and its properties is therefore key to designing suitable composites where the abrasion environment is known. Abrasive properties that have to be taken into account when designing composites for use in abrasive environments include: hardness, shape, size and ductility. When designing composites, the environment and conditions under which the abrasives function are also matters of consideration, and critical parameters of interest include: contacting areas, contact forces, velocities of interacting surfaces, the extent of lubrication, the temperature, and other ambient conditions (humidity, pressure). When designing composites for use in abrasive environments, appropriate design parameters are typically those that are interdependent on abrasive properties, and on the environment and conditions. Composite design parameters include: materials composition and properties, fractions of reinforcement to matrix, microstructures, surfaces (in particular topography and roughness), reinforcement/matrix interfaces, and the locations and orientations of reinforcements [5].

doi:10.1088/978-0-7503-3527-0ch1

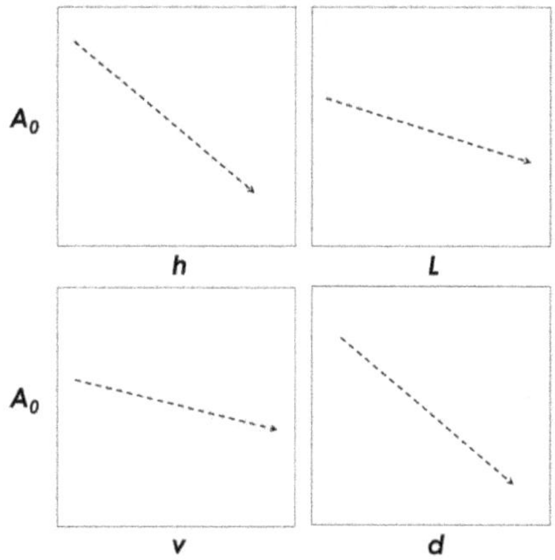

Figure 1.1. Specific wear rates in composites are 'generally' expected to decrease as the following parameters are increased: hardness (h), load (L), sliding velocity (v), and sliding distance (d).

Wear can be assessed in various ways. The classical method is to measure composite weight loss, however volume loss and optical methods (elucidating damage) are also mechanisms of assessment. Note that microscopy techniques are expected to be used alongside volume or weight loss measurements, to ensure the absence of foreign trapped particulates (e.g. those that may come from an abrasive). Nevertheless, since abrasion is also a function of dimensions and loading conditions, composite abrasion can be normalised to a *Specific Abrasion Rate* (also termed *Specific Wear Rate*), A_0 such that $A_0 = \frac{V}{L \cdot d}$, where V is the volume of composite abraded, L is the load applied during abrasion, and d is the distance abraded (sliding distance). The units of A_0 are $\frac{m^3}{Nm}$. This is a simpler approach than the *Archard Wear Law*, A_{archard} which incorporates the hardness, h, of the material such that $A_{\text{archard}} = \frac{V \cdot h}{L \cdot d}$ with units. Since composites are multi-component materials, it can sometimes be simpler to negate certain parameters that have a primarily localised influence to the properties. That said, abrasion in CMCs is typically less sensitive to localised hardness than in PMCs and MMCs. Perhaps the simplest way of assessing composite abrasion is by calculating an *Abrasion Rate*, A, which excludes the inversely proportional effect of load on wear. In such a case, $A = \frac{V}{d}$ and the units of abrasion rate are simply $\frac{m^3}{m}$.

It can be beneficial to know what to expect from an abrasion test as a first-order assessment of testing data. Figure 1.1 shows how A_0 is generally expected to decrease as a function of increasing parameters including: hardness (h), load (L), sliding velocity (v), and sliding distance (d).

1.2 Accelerated testing

Conditions affecting the performance of composites over their service life include but are not limited to: loading, temperature, moisture content, vibrations, and pressures.

These condition may occur in repetition or continuously, and engineers require predictive models to enable either the timely maintenance, or, the timely replacement of composite parts. Testing conducted over the entire service life of a composite product is time consuming and expensive, which makes incremental, iterative design difficult to achieve. To circumvent the problems associated with conducting long, expensive tests, engineers will test composites using one or more variables (e.g. temperature, loading) at accelerated rates (accelerated testing), by which means they are able to extrapolate from predictive models to enable incremental, iterative design. Accelerated tests are often categorised by their responses. A binary response (e.g. does the composite fail or not under a defined load) is termed an accelerated binary test (ABT). Responses that are related to the lifetime of the composite (e.g. fatigue, temperature, moisture) are accelerated life tests (ALTs). Degradation can occur through either physical or chemical means. Decomposition rates of composites (e.g. chemical or temperature degradation) are essentially accelerated degradation tests (ADTs). If only one degradation output variable can be measured, ADT is an accelerated destructive degradation test (ADDT), while if degradation outputs are sampled throughout the ADT, the ADT is termed an accelerated repeated measures degradation test (ARMDT). The following sections provide details of accelerated tests.

1.2.1 Accelerated life testing (ALT)

Cyclic loading can affect composites in a broad variety of applications, e.g. helicopter rotor blades, wind and tidal turbine blades, bridge loading, and fixed wing aircraft. Accelerated cyclic loading (fatigue testing) is commonly used to determine the fatigue life of composites and composite parts. In the real world, fatigue may be a slow process, as evidenced by the example of the service life a tidal turbine blade, which is in the order of 15–25 years. Testing the service life of a composite over this time span is unrealistic. To accelerate the assessment and to predict the fatigue life of a tested composite, loads are applied at high rates and recorded stresses are plotted against the number of cycles to failure. These are termed S–N curves, and are useful in predicting how long a composite will last as the number of cycles to fail at specific stresses can be identified. S–N curves of fibre reinforced plastics (FRPs) are typically different to those observed in engineering metals such as mild steel and aluminium, figure 1.2, and this is in part, due to the differing ways in which the materials fail.

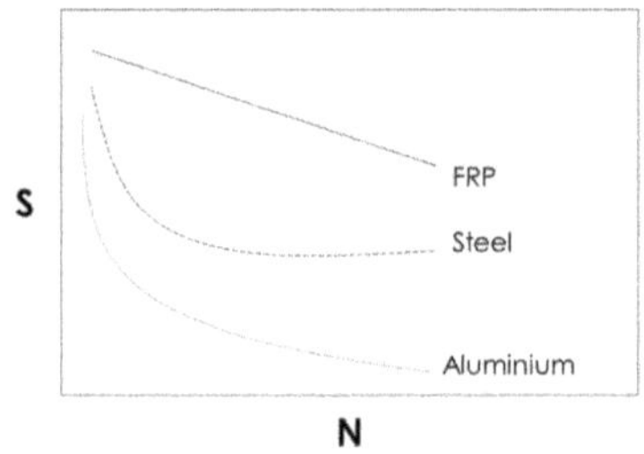

Figure 1.2. Comparison of S–N curves of fibre reinforced plastic (FRP) composite, mild steel, and aluminium. The abscissa axis in this plot is a logarithmic scale, while the ordinate axis is linear.

1.2.2 Accelerated degradation testing (ADT)

Environmental effects include the effects of temperature and moisture on the properties of FRP. A common method for accelerating temperature effects on FRP is to heat the composite above its working temperature but below its glass transition temperature (Tg), to artificially exacerbate the process of cross-linking and hence, polymer ageing within the composite structure. If heating is conducted above the matrix Tg, oxidation of the polymer might influence the properties of the composite and this needs to be avoided so that only the long term effects of temperature on the properties are expedited. Moisture ingress into composite is also a timely process and the rate at which moisture enters a composite can be accelerated through hygrothermal (or hydrothermal) ageing. This involves submersing composites for mechanical testing and/or characterisation (physical, chemical and/or optical) in heated water for period of time. It should be noted nevertheless, that hygrothermal ageing is a combined process of both moisture ingress and heat, and this needs to be taken into account when analysing data [6].

1.2.3 Time, temperature, stress superposition

When using elevated temperatures to accelerate the chemical, physical or mechanical effects of ageing in viscoelastic FRPs, analysis is often best done through the principles of superposition. Similarly, since the times of accelerated experimental tests are conveniently shorter than if testing were conducted over a composite's lifetime, both time and temperature superposition principles can be used to determine longer-term stress responses of heated (aged) composites. The time–temperature–stress superposition principle posits that the effect of a temperature or stress change on time-dependent material responses is equivalent to a uniform shift on a logarithmic scale [7]. Examples of time-dependent material responses include strain (elongation) and the modulus. Figure 1.3 shows an example of how the curve of a material response (in this case, strain) when plotted against the log of time will match over a wide range of temperatures.

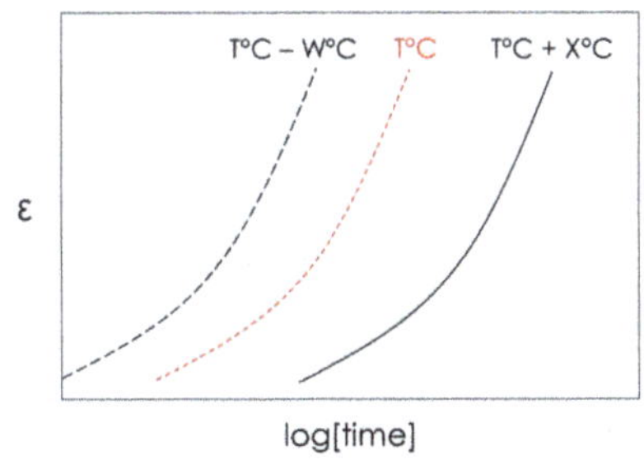

Figure 1.3. Strain plotted against log(time) curves at different temperatures. The time–temperature–stress superposition principle posits that these curves should match over a broad range of temperatures.

1.3 Additives

Additives used in composites are typically chemical or mineral addition agents applied to the matrix of a polymer matrix composite. In fewer instances, additives may be applied to fibres (e.g. sizing attached to fibre surfaces to improve adhesion at fibre–matrix interfaces). Additives used in composites can be categorised based on the purpose for their addition. There are three main groups: cost-saving additives, manufacture and process additives, and functional additives. In some cases there are cross-overs between these groups.

1.3.1 Cost-saving additives

Cost-saving additives are primarily used to decrease the costs of more expensive reinforcement and matrix materials and are thus also often termed *fillers*, *filling agents*, or *extenders*. The extent to which a filler affects the mechanical performance of the composite (either positively or detrimentally) depends on its strength of adhesion to the surrounding material, its intrinsic materials properties, its shape, the fraction of filler used, and the mode of loading it may be subjected to. The most common (and lowest cost) mineral filler is calcium carbonate followed by kaolin.

1.3.2 Manufacture and process additives

Manufacture and process additives are often employed to improve rheological performance, mould release, and dimensional stability. Mineral fillers for example, though usually considered cost-saving additives, often improve the dimensional stability of composites curing, as unlike matrix material, they do not shrink while composites are curing. As such, mineral fillers can be used to design for optimal processing where dimensional stability and reduced thermal expansion are desirable qualities during composite curing. Matrix micro-cracking in polyester matrix composites can be relieved during resin transfer moulding (RTM) processes through the addition of low profile additives (LPAs), thermoplastic particulates incorporated into unsaturated PE resins. Examples of LPAs include: poly vinyl acetates, poly methyl methacrylates, polystyrenes, and high density polyethylenes. LPAs consequently improve the surface finish of unsaturated polyester matrix composites by compensating for shrinkage. Dispersing agents (dispersant) are compounds used to enable the homogenous distribution of fillers or nano-particles within polymer matrices, thereby improving both the local and global mechanical properties of particle composites. Improved dispersion also gives rise to easier processing and superior opacity/consistency. Dispersing agents are commonly used in composite paints and coatings where agglomeration can be a problem. Nano-cellulose hornification (agglomeration) in cellulose nanocomposites have been shown to benefit from the use of, e.g. glucomannan and xylan dispersant.

1.3.3 Functional (performance) additives

Functional (or performance) additives can be incorporated into the composites design process, adding value through function to the final composite products. For example, while the detrimental effects of filler on the mechanical properties of composites are often cited, fillers can in fact improve mechanical performance in fire by decreasing the fraction of easily burned organic matrix material. Mineral fillers can also decrease the uptake of water into the composite, and can thus be incorporated into the composite design to circumvent moisture ingress when this quality is of high priority. Mineral fillers such as alumina tetrahydride (ATH) (a water releasing filler under high temperatures) and calcium sulphate (water releasing filler at low temperatures) can be used to design for fire resistance in composites. Other fire retardant additives include bromine, chlorine, and phosphorous.

While some functional additives decrease matrix viscosity to improve reinforcement wetting, on occasion thixotropic additives may be added to composites in their uncured state to increase viscosity. Thixotropes (e.g. fumed silica or kaolinites) allow matrices to retain high viscosity when at rest. This is useful in certain applications where resin flow is to be avoided, e.g. high matrix fraction glass fibre composites applied to the vertical surfaces or bottom up applications such as the underside repair of fractured timber beams.

Sizing applied to fibre surfaces are functional additives designed to improve fibre–matrix adhesion. Glass fibres are typically coated with silane based sizing, while carbon fibres are often coated with epoxy based sizing. The intermolecular interactions between fibre, sizing and matrix, as well as the final sizing topography are the primary drivers behind fibre–matrix adhesion.

Functional additives can also be external to the composite part, e.g. mould release agents applied to the surface of the part mould. Gel coats are polyester resin based finishing agents applied to the surface of the mould, which then attaches through the surface of the cured composite part or product (e.g. boat hulls). These coats have various functions including: improved weathering, improved UV protection, improved thermal protection, and improved abrasion resistance. Gel coats typically include thixotropes so as to retain high gel viscosities and to decrease sag. Corrosion resistance and weather protection is also possible through the use of nylon or polyester based surface veils. Though these are not small-sized additive compounds, they have a performance function and are essentially *add-ons* to cured composite parts, and as such are included herein.

1.4 Adhesion

There are three fundamental forms of adhesion. *Specific adhesion*, which is a phenomenon influenced by molecular interactions at different surfaces, *mechanical adhesion*, which is a topo-geometry driven adhesion, and *effective adhesion*, which is the coupling between specific and mechanical adhesion. Specific adhesion mechanisms include: diffusion theory, absorption mechanisms, chemical reaction theories, electrostatic interactions, and acid-base reaction theories of adhesion. Mechanical adhesion is a result of interlocking. Knowledge on the individual theories will not be covered here since these more can be more appropriately discussed in the section on Wetting.

In composites engineering, adhesion can divided into two distinct areas: adhesion *in* composites, and the adhesion *of* composites.

1.4.1 Adhesion *in* composites

Adhesion in MMCs, CMCs and FRPs is most typically understood to occur between reinforcement and matrix. In fact, very thin layers of coating (additives) are often applied to reinforcement surfaces. In MMCs the purpose of a reinforcement coating is usually to avoid chemical reaction with the matrix. Carbon fibre reinforcements in aluminium MMCs are for example, coated with nickel or titanium bromide to prevent the formation of aluminium carbide at the surface of the carbon fibres. Ceramic-reinforced ceramic CMCs do not usually need surface preparation since the manufacturing processes guide adhesion, however, well dispersed reinforcements (using dispersing agents) tend to improve adhesion between the composite components. In FRPs fibres are sized (coated) with, e.g. silanes (glass fibres) or epoxides (carbon fibres) to improve adhesion strength between fibre and matrix. Sizing is an example of *effective adhesion* as topo-geometry couples to molecular interactions to enable fibre–matrix adhesion. Methods for measuring fibre bond strength *in* composites include:

- **Micro-bond**: These tests involve pulling a single reinforcing fibre out of a bead (or droplet) of cured matrix, usually through a knife edge slit. The critical properties leading to interfacial fracture are derived from the force required to pull the fibre out. Two criticisms of the micro-bond test include (i) results are highly scattered (i.e. not consistent) and (ii) whether the test elucidates true interfacial properties is questionable, since micro-bond test results are often inconsistent with those determined using other test methods [8].
- **Micro-debond**: Micro-debonding aims to create debonding cracks at the fibre–matrix interface. It achieves this by applying pressure to the polished surface of a fibre embedded in matrix. The load is applied in the axial direction of the fibre such that interfacial shear is the prime failure mode. The debonding force is the load measured at the point of debonding between fibre and matrix [9].
- **Single fibre pull out**: This test involves applying a tensile pull to single fibres embedded within a matrix material. Critically, the length of fibre embedded into matrix should not exceed a limiting length at which catastrophic failure ceases to occur (i.e. complete debonding of the entire fibre). The basis equation for the calculation of single fibre pull out fracture energies below this limiting length, is $G = (P^2/2\pi d)(dC/da)$, where G is the fracture energy for crack opening (mode I), P is the tensile load applied to the fibre, d is the diameter of the fibre, and dC/da is the compliance of the structure as a function of crack length a. Many mode I models for single fibre pull out are modified from this basis model, however, interfacial fracture during a fibre pull out is predominantly in shear (mode II). The basis for calculating the maximum interfacial bond strength in shear, τ_{bond}, is $\tau_{\text{bond}} = \frac{P_{\text{init}}}{\pi dl}$, where P_{init} is the initial or interfacial fracture load, d is the fibre diameter and l is the length of fibre embedding. Interfacial debonding in shear, τ_{debond}, can be calculated as, $\tau_{\text{debond}} = \frac{dP/da}{1/\pi d}$, while shear sliding, τ_{sliding}, can be calculated as $\tau_{\text{sliding}} = \frac{dP/d\delta}{1/\pi d}$, where $dP/d\delta$ is the gradient of the pull-out tail from the load–displacement curve [10].

- **Single fibre push out**: The fundamental set up for a single fibre push out test is in principle similar to a micro-debond, however the purpose and approach to analysis is different. Whereas in a micro-debond test the point of failure is the measurement value, a single fibre push out test takes measurements through the entire loading cycle of the test. This means that not only is the initial peak strength recorded, but also, the elastic loading, progressive debonding, interfacial sliding and fibre push through. The curve is essentially a stress–strain (or load–displacement) curve that rises in load at its end as a result of indenter collision against the matrix.

1.4.2 Adhesion *of* composites

The adhesion of composites simplifies the manufacturing process of large or geometrically complex parts. This results in the reduction of costs and a broader application range for composite materials. The adhesive bonding of composites offers numerous other benefits including: enabling large areas for stress bearing between adherends, good shock absorption, excellent strength to weight ratio, the connecting of dissimilar composites (e.g. GFRP with CFRP), and an ability to seal. Indubitably, there will also be disadvantages to bonding composites including: surfaces may need to be prepared prior to application (e.g. using coupling agents, primers or adhesion promoters), extra curing requirements in at least the bonded areas, pressure application in bonded areas, ageing of adhesives, and quality inspection of bonding is difficult using NDT methods.

Common adhesively bonded composite joints include: lap joints, scarf joints, strap joints, tapered strap joints, butt joints, stepped lap joints (e.g. ply drops), L-joints, T-joints, wrapped FRP joints, flitching, and moment resisting plates. Sandwich panels may be joined using flat panel edging, flat panel joining and extruded joining methods. Common test methods for composite joints include single lap shear, double shear, joggle lap shear, reverse bent lap joint, joint peel, and double cantilever beam.

Common adhesives that work through chemical reaction during the transition from resin to solid include: epoxies, toughened acrylics, anaerobics, polyurethanes, cyanoacrylates, silicones, phenolics and polyamides. Common adhesives that undergo a physical change during the transition from resin to solid include: polyvinylacetates, rubbers, hot-melts, tapes and plastisols. Chemically reactive adhesives form far stronger bonds than adhesives that undergo physical change. Selecting adhesives is typically broken down into three principles: type of joint, materials to be joined, function of joint, properties of adhesives, service life conditions, issues related to manufacturing, and overall cost.

1.5 Air bubble voids

An air-bubble void (ABV) is, essentially, air that has been entrapped within a composite. These voids are detrimental to mechanical integrity and their presence increases the statistical likelihood of early composite failure. Air-bubble voids are found between reinforcement plies and within bond-lines. They may also be confined to localised non-interconnected areas within the composite.

Voids appear as a result of manufacturing, with certain fibre reinforced plastic processing methods yielding higher fractions of voids than others. The differences are related to variations in both thermodynamics and rheology between the processes. In liquid composite moulding (LCM) methods such as resin transfer moulding (RTM) and vacuum-assisted resin transfer moulding (VARTM), air entrapment during resin flow is a primary cause of ABV formation. Secondary causes of ABV formation include volatile gases created during the curing process, and the nucleation of dissolved gases within the resin matrix. Though the mechanisms are not fully understood, air entrapment occurs as a result of inhomogeneities of the fibre architectures essentially 'blocking' gas permeability, and creating an environment for differential rates of resin flow that are confined to local regions within the architecture. Modern methods including out-of-autoclave (OoA) processing, and automated prepreg laying (APL) have high accuracy, require low times and are low cost methods. A drawback of using OoA and APL methods nevertheless, is voidage.

Voids form at macro, meso and micro length scales. Both meso and micro scale voids can form through the incomplete impregnation of resin fibres inside and between tows. Meso-voids tend to form between tows as the space available is typically 'large' and viscous flow of the resin is driven by hydrodynamic forces. Micro-voids tend to form between individual fibres and inside small gaps within fibre tows. In these geometrically tighter environments, surface tension dominates and capillary pressures drive resin flow (wicking). Voids in this environment can be at the meso-scale if viscous flow predominates over capillary driven flow. As a general rule, meso-voids decrease as a function of an increasing capillary number, while micro-voids increase as the capillary number rises [11]. An important design parameter to minimise voids can therefore be the capillary number, which can be deduced from the intersection of the meso and micro curves in a chart connecting void content to a modified capillary number, $Ca = \frac{\mu V}{\gamma \cos\theta}$ [12], figure 1.4; where μ is the dynamic viscosity of the resin, V is its flow velocity, γ is the resin surface tension, and $\cos\theta$ is the resin-fibre contact angle.

Macro-voids form during LCM for three predominate reasons. These include: (i) resin pressures around dry regions being too low to allow shrinkage or collapse of the voids, or forced movement towards LCM vents (ii) resin flow reaching LCM

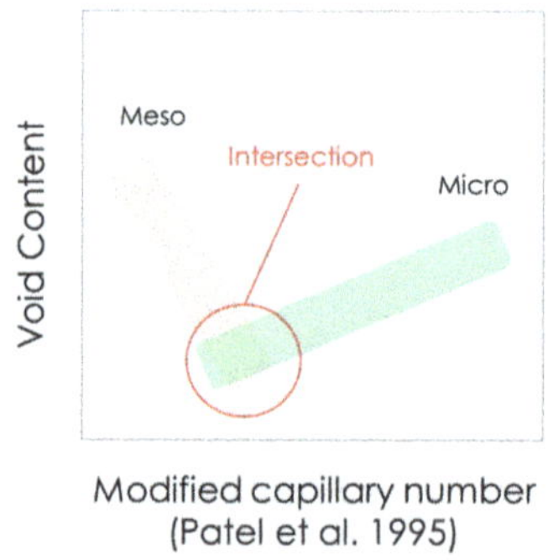

Figure 1.4. Intersection between meso and micro curves in a void-Ca plot as a design parameter for reduced voidage.

vents before completely impregnating and wetting the LCM preform and (iii) the presence of air within dry regions arising through an incomplete vacuum.

In prepreg technologies, voids form during the curing stages of the prepregs. This can be due to the physical entrapment of air prior to curing, or due to the evolution of volatile gases during curing. Though void formation is less common using autoclave processing methods as compared to LCM, similarly to prepreg technologies, voids can also form through the release of volatile gases during curing. Voids can be collapsed to some extent by judiciously controlling pressure and temperature, however voidage can also be avoided by controlling the physical properties of the resin, such as viscosity (lower viscosity typically resulting in lower voidage).

Moisture absorption, which in itself affects the mechanical performance of composites, is also exacerbated by the presence of voids. While Fickian behaviour is typically observed in fibre reinforced plastics [6], high voidage results in a shift from Fickian to anomalous-Fickian behaviour. Both voidage and water ingress are already understood as detrimental to the mechanical properties of composites. Combining the two naturally decreases mechanical integrity, negatively affecting the longevity of composites.

1.6 Angle-ply laminate

In composite laminate theory (CLT), ply stiffness can be assumed to be under a state of plane stress, equation (1.1), where Q_{ij} is termed the reduced stiffness, and σ_1 and σ_2 are normal stresses, σ_6 is shear stress, ε_1 and ε_2 are normal strains, and ε_6 is shear strain. Q_{ij} is related to stiffness, C_{ij} as $Q_{ij} = C_{ij} - \frac{C_{i3}C_{j3}}{C_{33}}(i, j = 1, 2, 6)$. σ for a 0° lamina can be calculated thence as $\sigma_i = Q_{ij}\varepsilon_j$.

$$\begin{Bmatrix} \sigma_1 \\ \sigma_2 \\ \sigma_6 \end{Bmatrix} = \begin{bmatrix} Q_{11} & Q_{12} & 0 \\ Q_{21} & Q_{22} & 0 \\ 0 & 0 & Q_{33} \end{bmatrix} \begin{Bmatrix} \varepsilon_1 \\ \varepsilon_2 \\ \varepsilon_6 \end{Bmatrix} \tag{1.1}$$

Q_{ij} relates to engineering constants as: $Q_{11} = \frac{E_{11}}{1 - \nu_{12}\nu_{21}}$; $Q_{12} = \nu_{21}Q_{11}$; $Q_{22} = \frac{E_{22}}{1 - \nu_{12}\nu_{21}}$; $Q_{16} = 0$; $Q_{26} = 0$; $Q_{66} - G_{12}$, where E is elastic modulus, ν is Poisson's ratio, and G is shear modulus.

Ply angles in laminates are often varied. Calculating the reduced stiffness so as to take into account the ply angle, $\bar{Q}_{ij}$, requires consideration of the effects of these angles on the properties, equation (1.2), where $m = \cos(\theta)$ and $n = \sin(\theta)$, θ being the ply orientation angle.

$$\begin{Bmatrix} \bar{Q}_{11} \\ \bar{Q}_{22} \\ \bar{Q}_{12} \\ \bar{Q}_{66} \\ \bar{Q}_{16} \\ \bar{Q}_{26} \end{Bmatrix} = \begin{bmatrix} m^4 & n^4 & 2m^2n^2 & 4m^2n^2 \\ n^4 & m^4 & 2m^2n^2 & 4m^2n^2 \\ m^2n^2 & m^2n^2 & m^4 + n^4 & -4m^2n^2 \\ m^2n^2 & m^2n^2 & -2m^2n^2 & (m^2 - n^2)^2 \\ mn^3 & -mn^3 & n^3m - m^3n & 2(n^3m - m^3n) \\ mn^3 & -m^3n & m^3n - n^3m & 2(m^3n - n^3m) \end{bmatrix} \begin{Bmatrix} Q_{11} \\ Q_{22} \\ Q_{12} \\ Q_{66} \end{Bmatrix} \tag{1.2}$$

$\bar{Q}_{ij}$ can then be used to determine the extensional stiffness (A_{ij}), equation (1.3), the coupling stiffness (B_{ij}), equation (1.5), and the bending stiffness (D_{ij}), equation (1.4), where t_k is the thickness of the kth lamina and $\bar{z}_k$ is the distance from the mid-plan to the centroid of the kth lamina.

$$A_{ij} = \sum_{k=1}^{n} \left(\bar{Q}_{ij}\right)_k t_k \tag{1.3}$$

$$B_{ij} = \frac{1}{2} \sum_{k=1}^{n} \left(\bar{Q}_{ij}\right)_k t_k \bar{z}_k \tag{1.4}$$

$$D_{ij} = \frac{1}{3} \sum_{k=1}^{n} \left(\bar{Q}_{ij}\right)_k \left(t_k \bar{z}_k^{-2} + \frac{t_k^3}{12} \right) \tag{1.5}$$

These in turn can then be used to calculate the forces, N, and moments, M, according to equation (1.6), which is shown in the expanded form in equation (1.7)

$$\begin{Bmatrix} N \\ M \end{Bmatrix} = \begin{bmatrix} A & B \\ B & D \end{bmatrix} \begin{Bmatrix} \varepsilon^0 \\ \kappa \end{Bmatrix} \tag{1.6}$$

$$\begin{Bmatrix} N_{xx} \\ N_{yy} \\ N_{xy} \\ M_{xx} \\ M_{yy} \\ M_{xy} \end{Bmatrix} = \begin{bmatrix} A_{11} & A_{12} & A_{16} & B_{11} & B_{12} & B_{16} \\ A_{12} & A_{22} & A_{26} & B_{12} & B_{22} & B_{26} \\ A_{16} & A_{26} & A_{66} & B_{16} & B_{26} & B_{66} \\ B_{11} & B_{12} & B_{16} & D_{11} & D_{12} & D_{16} \\ B_{12} & B_{22} & B_{26} & D_{12} & D_{22} & D_{26} \\ B_{16} & B_{26} & B_{66} & D_{16} & D_{26} & D_{66} \end{bmatrix} \begin{Bmatrix} \varepsilon_{xx}^0 \\ \varepsilon_{yy}^0 \\ \varepsilon_{xy}^0 \\ \kappa_{xx} \\ \kappa_{yy} \\ \kappa_{xy} \end{Bmatrix} \tag{1.7}$$

Here, κ_x, κ_y and κ_{xy} are the laminate curvatures calculated as $\frac{1}{r}$, where r is the radius of curvature. An angle-ply laminate has equal numbers of plies of the same thickness and material that are oriented as $+\theta$ and $-\theta$, e.g. [+45/ − 45/ − 30/ + 30]. In angle-ply laminates $A_{16} = A_{26} = 0$ because mn is the same for $\bar{Q}_{16} = -\bar{Q}_{16}$ and $\bar{Q}_{26} = -\bar{Q}_{26}$. As such, angle-ply laminates are considered specially orthotropic.

1.7 Anisotropy

A material has anisotropy, or is anisotropic, when it exhibits variations in physical properties in different directions. Anisotropy in composites can be complex to define, since its interpretation depends on the length scale at which the observer intends to define. A homogenously polydispersed particulate reinforced composite for example, may exhibit isotropic mechanical properties in orthogonal axes at a centimetre length scale. However at a smaller length scale (e.g. nano/micro), the same composite will exhibit a range of mechanical properties both along specific axes, and over different axes. Engineering composites are usually, for this reason, characterised for isotropy and anisotropy at the larger length scales as these are the scales at which they are used.

At the higher length scales, many composites are anisotropic. Anisotropic elasticity (linear) is often understood through vector–tensor relationships as this is mathematically simple and enables, e.g. the modelling of mechanical behaviour in composites. Hooke's law (generalised), where normal stresses σ_1, σ_2, σ_3 and shear stresses σ_4, σ_5, σ_6 have a linear dependence via a matrix of 36 stiffnesses, $C_{i,j}$, (the stiffness matrix) on normal strain components ε_1, ε_2, ε_3 and shear strain components ε_4, ε_5, ε_6, respectively, equation (1.8).

$$\begin{Bmatrix} \sigma_1 \\ \sigma_2 \\ \sigma_3 \\ \sigma_4 \\ \sigma_5 \\ \sigma_6 \end{Bmatrix} = \begin{bmatrix} C_{11} & C_{12} & C_{13} & C_{14} & C_{15} & C_{16} \\ C_{21} & C_{22} & C_{23} & C_{24} & C_{25} & C_{26} \\ C_{31} & C_{32} & C_{33} & C_{34} & C_{35} & C_{36} \\ C_{41} & C_{42} & C_{43} & C_{44} & C_{45} & C_{46} \\ C_{51} & C_{52} & C_{53} & C_{54} & C_{55} & C_{56} \\ C_{61} & C_{62} & C_{63} & C_{64} & C_{65} & C_{66} \end{bmatrix} \begin{Bmatrix} \varepsilon_1 \\ \varepsilon_2 \\ \varepsilon_3 \\ \varepsilon_4 \\ \varepsilon_5 \\ \varepsilon_6 \end{Bmatrix} \tag{1.8}$$

In reality, the stiffness matrix is symmetric and as such there are only 21 independent elastic constants, equation (1.9)

$$\begin{Bmatrix} \sigma_1 \\ \sigma_2 \\ \sigma_3 \\ \sigma_4 \\ \sigma_5 \\ \sigma_6 \end{Bmatrix} = \begin{bmatrix} C_{11} & C_{12} & C_{13} & C_{14} & C_{15} & C_{16} \\ & C_{22} & C_{23} & C_{24} & C_{25} & C_{26} \\ & & C_{33} & C_{34} & C_{35} & C_{36} \\ & & & C_{44} & C_{45} & C_{46} \\ & & & & C_{55} & C_{56} \\ & & & & & C_{66} \end{bmatrix} \begin{Bmatrix} \varepsilon_1 \\ \varepsilon_2 \\ \varepsilon_3 \\ \varepsilon_4 \\ \varepsilon_5 \\ \varepsilon_6 \end{Bmatrix} \tag{1.9}$$

Equation (1.9) can also be inverted to express strains as linearly dependent on stress, this time via a compliance matrix, equation (1.10).

$$\begin{Bmatrix} \varepsilon_1 \\ \varepsilon_2 \\ \varepsilon_3 \\ \varepsilon_4 \\ \varepsilon_5 \\ \varepsilon_6 \end{Bmatrix} = \begin{bmatrix} S_{11} & S_{12} & S_{13} & S_{14} & S_{15} & S_{16} \\ & S_{22} & S_{23} & S_{24} & S_{25} & S_{26} \\ & & S_{33} & S_{34} & S_{35} & S_{36} \\ & & & S_{44} & S_{45} & S_{46} \\ & & & & S_{55} & S_{56} \\ & & & & & S_{66} \end{bmatrix} \begin{Bmatrix} \sigma_1 \\ \sigma_2 \\ \sigma_3 \\ \sigma_4 \\ \sigma_5 \\ \sigma_6 \end{Bmatrix} \tag{1.10}$$

Equations (1.9) and (1.10) are simplifications of reality as it is very difficult to mathematically capture true anisotropic behaviour. Nevertheless, composites often exhibit a degree of material symmetry that allows for further mathematical simplification. These behaviours are specialised forms of elastic anisotropy and include:

- Orthotropy: a composite with three orthogonal planes of symmetry.
- Transverse isotropy: a composite with a primary directional property orthogonal to which the composite exhibits isotropy.

1.8 Areal weight

The areal weight, A_w, of tape, fabric or composite is fundamentally the weight of fibre per unit area such that $A_w = w \times l$, where w is the width and l is the length of the tape, fabric or composite. A_w is usually expressed in grams per square metre, or gsm.

The fibre areal weight, A_F, refers to the weight of fibre. This is different to the prepreg areal weight, A_P, which includes the weight of pre-impregnated resin. The fibre areal weight of fabrics may be calculated according to equation (1.11), where N_f is the number of filaments in each tow, N_t is the number of tows per unit width of fabric, r_f is the radius of the fibre cross section, and ρ_f is fibre density [13].

$$A_F = 2N_f N_t r_f \rho_f \tag{1.11}$$

Crimp angle can affect A_F. An A_F increase of 1%, 3% and 6.5% can be expected to occur at crimp angles of 10°, 20° and 30°, respectively.

A_F is a useful parameter as regardless of the extent of bleeding, air-bubble void formation, or resin that is confined within the final cured composite, the fibre areal weight remains the same. Cured ply thickness, C_{pt}. for example, can be crudely estimated according to equation (1.12), where V_f is the fibre volume fraction.

$$C_{pt} = \frac{A_F}{\rho_f V_f} \tag{1.12}$$

The matrix mass fraction, m_{mat}, can also be deduced from A_F in accordance with equation (1.13), where A is the ply area and m_{cured} is the cured mass of composite.

$$m_{mat} = 1 - \frac{A_F \times A}{m_{cured}} \tag{1.13}$$

The fraction m_{mat} can also be approximated in prepreg materials if A_F within A_P (prepreg areal weight) is known, equation (1.14).

$$m_{mat} = 1 - \frac{A_F}{A_P} \tag{1.14}$$

The areal weight of composites can sometimes be used as 'very general' indicators for certain mechanical and dynamic mechanical properties. The ballistic strength of composites, for example, has been considered as a property that increases linearly with areal weight [14] in woven fabric composites incorporated in a continuum matrix. Both modulus and strength of FRPs with differing orientation are properties that show the areal weight has an optimum that itself depends on the type of fibre and its physical properties, the orientation of fibre placement, and the mode of loading the composites are subjected to [15].

1.9 Aramid

Aramid is a synthetic organic polymer used as a high-temperature, high-tenacity reinforcement in advanced composites. Trade name examples of aramid are Kevlar,

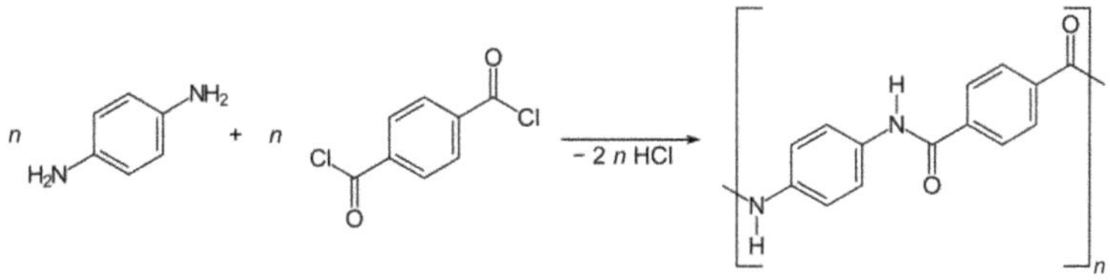

Figure 1.5. The reaction of 1,4-phenylene-diamine (para-phenylenediamine) with terephthaloyl chloride yielding the aramids Kevlar and Twaron. Wikipedia: Roland.chem (CC0 1.0).

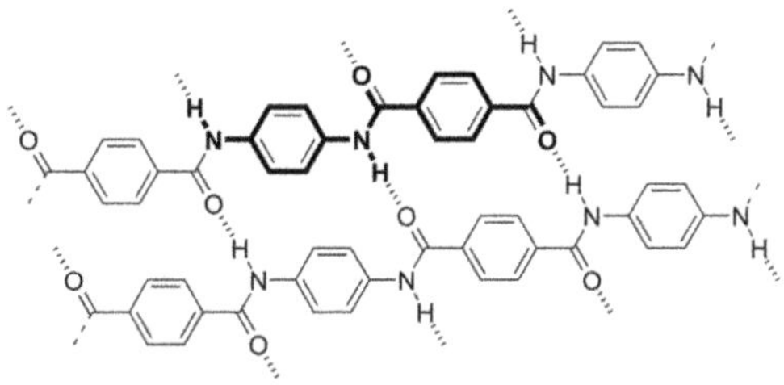

Figure 1.6. Hydrogen bonding between aramid molecules resulting in single axis oriented and bound molecular structures found in para-aramids such as Twaron and Kevlar. Wikipedia: Hbf878 (CC0 1.0).

Nomex and Technora (DuPont), and Twaron (Teijin). Aramid is most usually used in the forms of fibre or yarn, with the best adhering matrix for aramid being epoxy resin, though vinyl esters and isophthalic polyesters (not orthophthalic polyesters) are also good matrix materials in aramid fibre composites. Aramids are chemically prepared by reacting amine with carboxylic acid halide groups. Figure 1.5 shows an example reaction between 1,4-phenylene-diamine (para-phenylenediamine) and terephthaloyl chloride, which results in aramid like Kevlar and Twaron. As can be seen in this figure, aramids are linear backbones molecules with the potential for hydrogen bonding (electrostatic interactions) between proximal $N - H$ and $=O$. The interaction between these along entire molecular chains results in high levels of hydrogen bonding binding the molecules together along a single axis, figure 1.6 and creating crystallinity. The large scale of intermolecular hydrogen bonding decreases free volumes between molecules and ensures the covalent molecular backbones of aramids are structurally beneficial, with little possibility for intermolecular shearing when loaded in tension. Since these molecules are geometrically speaking, long and thin, they tend to work well in tension, but are not as effective in compression, where molecules may buckle and lose the intermolecular strength of binding provided by the hydrogen bonds.

There are two major classes of aramids that differ not only by their mechanism of manufacture, but also by their final properties and use. These classes are meta-aramids and para-aramids.

1.9.1 Meta-aramids (m-aramids)

M-aramid fibres are manufactured by wet-spinning in a chemical solution. The resulting material is a semi-crystalline fibre with partial molecular chain orientation along its fibre axis. A unique property of meta-aramid fibres is that they are able to withstand tensile stress while subjected to temperatures (up to 400 ° C), as well as to naked flames. M-aramids are generally considered temperature, abrasion and chemically resistant fibres. These fibres are used in applications such as: heat resistant clothing for fire-fighters, military personnel and other first responders, heat shielding applications in automotive and aerospace industries, and as insulation material. Commercial m-aramid fibres include: **Nomex** produced by DuPont Co., **Teijinconex** produced by Teijin Ltd., and **Fenilin** produced in Russia. Each of these m-aramids are essentially poly(m-phenylene isophthalamide).

1.9.2 Para-aramids (p-aramids)

P-aramid fibres are manufactured by a dry jet, wet-spinning method that results in fibres with high levels of crystallinity. Due to their high crystallinity, p-aramid fibres are used in high stiffness and high strength applications such as: body armour (ballistics protection), as FRP composite reinforcing, in race cars, tapes and webbing, ropes and cables, friction materials, and as rubber reinforcements. Commercial p-aramid fibres include: **Kevlar** produced by DuPont Co. and **Twaron** produced by Akzo Nobel—both of which are poly(p-phenylene terephthalamide), and **SVM** Poly(5-amino-2-(p-aminophenyl) benzimidazole terephthalamide) produced in Russia.

1.9.3 Comparison of m-aramid and p-aramid properties

See table 1.1.

1.10 Aspect ratio

Aspect ratio, A_{ratio}, or aspect, is defined in FRPs and PRCs as $A_{\text{ratio}} = \frac{l}{d}$, and in SCs as either $A_{\text{ratio}} = \frac{l}{d}$ or $A_{\text{ratio}} = \frac{l}{t}$, where l is the reinforcement length, d is the reinforcement diameter, and t is the reinforcement thickness. In cases of, e.g. blocky particulates in PRCs, or differential thickness/width reinforcements in SCs, d or t can be approximated by averaging across dimensions.

Aspect is a vital parameter in composites design, as it is a means to defining the effectiveness of the reinforcement under different conditions of loading.

Table 1.1. Comparison of m-aramid and p-aramid properties. Reprinted from [16] (CC BY 3.0).

Fibre class	p-aramid	p-aramid	p-aramid	m-aramid	m-aramid
Trade name	Kevlar 129	Kevlar 149	Twaron	Nomex	Kermel
Density (g cc^{-1})	1.44	1.47	1.44	1.38	1.34
Tenacity (N/tex)	2.35	1.7	1.95	0.4	0.36
Tensile strength (GPa)	3.32	2.40	2.74	0.55	0.48
Tensile modulus (GPa)	75	160	95	17	4.2
Elongation at break (%)	3.6	1.5	3.3	35	16

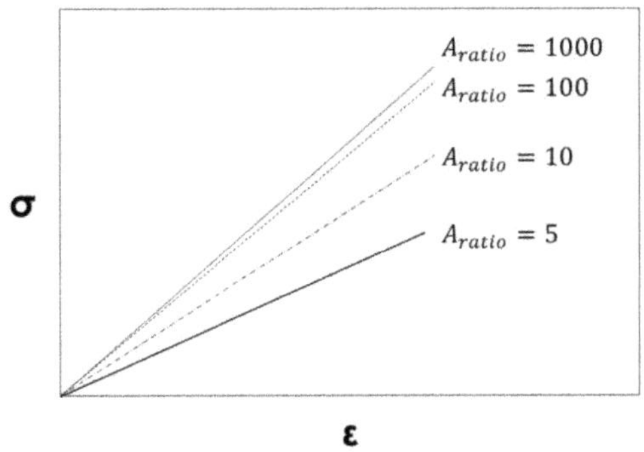

Figure 1.7. σ–ε curves showing the effect of aspect ratio of unidirectional FRPs subjected to a tensile pull in the direction of the fibre axis.

A material-dependent ratio-based limitation exists for reinforcement to be used to its fullest potential. A cubical (blocky) reinforcement, for example, if subjected to a tensile pull along its axis, will generally fail at its interface before it starts to experience damage (i.e. it remains in its recoverable elastic state). If this cubical reinforcement is lengthened along its axis (also the direction of tension), there is a greater surface area of adhesion and at a critical length, the bond stress in shear ($\tau_{\text{interfacial}}$) equals the yield or fracture strength of the reinforcement. This critical length is termed the *critical aspect ratio*, S_c and is the optimal aspect ratio for reinforcing. $S_c = \frac{\sigma_t}{2\tau_{\text{interfacial}}}$. It should be noted that this is based on (a) the understanding of elastic stress transfer, σ_r, such that $\sigma_r = E_r\varepsilon_r[1 - \cosh(\frac{nx}{r})\text{sech}(nA_{\text{ratio}})]$, where E_r is the Young's modulus of the reinforcement, ε_r is the applied strain in the composite, n is a dimensionless constant such that $n = \left[\frac{2E_m}{E_r(1+\nu_m)\ln(\frac{1}{V_r})}\right]^{\frac{1}{2}}$, E_m is the Young's modulus of the composite matrix, ν_m is the Poisson's ratio of the matrix material, and x is the axial distance from the midpoint of the reinforcement, and (b) understanding that the variation of bond stress in shear along the surface of the reinforcement is $\tau_{\text{interfacial}} = \frac{n\varepsilon_r}{2}E_r \sinh(\frac{nx}{r}\text{sech}(ns))$, where r is the radius or averaged length from cross-section centroid to reinforcement edge. The elastic theory behind A_{ratio} and its effect on mechanical properties of unidirectional FRPs is illustrated in figure 1.7, the generic trend here being that higher aspect ratios improve the Young's modulus when subjected to a tensile pull in the direction of the fibre axis.

The above is true for reinforcement (e.g. fibre) in tension. Reinforcements may also be subjected to, e.g. compressive, bending, shear, torsion, etc. For example, reinforcements cannot have an equivalent stress transfer aspect ratio in tension and shear, since in tension, this ratio is high, which in compression would result in reinforcement buckling. This would in turn impose bending strains and lateral shear across the width of the fibre. Thus, in the case of compression, reinforcement may in fact need to be decreased in its aspect in order to improve its ability to deform in linearly along its axis without bending.

1.11 Autoclave

Autoclaves are closed pressure vessels used to manufacture composite parts and materials. Uncured composite parts can be moulded in an autoclave, which is typically heated under pressure using inert gas (e.g. CO_2 or N_3), thereby enabling the transfer of heat and pressure to a composite component to consolidate it firmly and to allow it to cure. The process bears some similarities with hot pressing, the main difference being how pressure and heat are applied. Pressure is important in an autoclave, as it decreases porosity (e.g. air-bubble voids), retains shape around the mould, and allows for improved control of the fibre fraction of composites. Isostatic pressures can be applied to composite material around a mould by vacuum bagging the component prior to pressure/temperature application. Under pressure, the bag compacts the composite material and the intimacy of contact between composite and mould will therefore depend on the magnitude of pressure applied. Bleeders (absorbent materials) are used to channel out excess air and to soak up excess resin when curing under pressure and temperature. Prepreg laminates applied to the mould prior to vacuum bagging are covered in peel-ply as this enables the manufacture of composite products with higher quality surface finishes. Designers tend to adjust the autoclave moulding process to maximise air expulsion, while minimising excess resin flow. Curing pressures are typically in the range of 3–12 MPa [17], and prepregs tend to offer superior processability. The fundamental set-up for autoclave moulding is shown in figure 1.8.

Products formed through autoclave moulding are normally high quality and high density, and the process is thus one that is used regularly by, e.g. aerospace industries using carbon or glass FRPs. Since pressure vessel technologies are well established technologies, advancements in the sizes of autoclave are enablers for the manufacture of extremely large composite parts. ASC Process Systems (USA) for example,

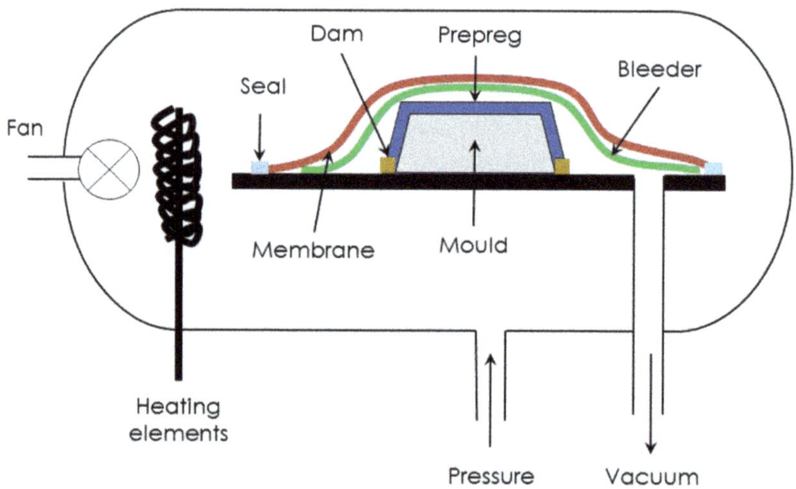

Figure 1.8. Fundamental set-up for autoclave moulding.

were responsible for building an autoclave for Boeing Charleston with an internal volume of 2321 m^3, to cure fuselage sections (232° C/1 MPa) for the Boeing 787 Dreamliner.

Autoclave manufacture is essentially a two stage process. In the first stage, vacuum and pressure are applied while the desired temperature is reached. The ramped heating allows resin viscosity to decrease gradually and allows for the release of volatiles and air bubbles. The reduced viscosity of the resin concurrently improves wetting of fibres and improves flow, which in turn helps with the movement of volatiles and air bubbles. In the second stage, the temperature is ramped to the final curing temperature, after which resin viscosity stabilises and the material starts to cure. In thermoset resins, curing is a chemical process that results in cross-linking and as such the curing cycle is most important in thermoset matrix composites. Thermoplastic matrix composites are autoclave moulded at high temperatures and bagging techniques used for thermoset matrix composites are adapted to suit [18]. Though autoclave moulding has several benefits previously mentioned, thermoregulation is one major weakness of the process and results in non-uniform part-cure. This is partly a function of the shape of the mould, and partly because traditional autoclaves pump heat from heating elements using fan-based technologies. Modern control systems use sensor-driven technologies to measure and regulate uniform temperatures in autoclaves.

1.12 Automated material placement (AMP)

AMP technology has been an enabler for significant advances in the aerospace industries, e.g. by influencing weight reductions in civil aircraft. Boeing for example, built much of its 787 using carbon fibre reinforced plastic (CFRP). By replacing aluminium with CFRP, Boeing was able to reduce weight, improve fatigue performance, improve corrosion resistance, and reduce the 787's overall fuel consumption.

Fibre reinforced plastics are often manufactured by laying individual thin plies of prepreg. As such, accuracy, repeatability, ease in controlling ply orientation, and speed of manufacture have been drivers for the development high quality automation technologies. Currently AMP technologies are up to three times more precise in maintaining positional tolerances than hand laid plies, and are in many cases, >10 times faster.

AMP machines for prepreg tows are available as filament winding (FW) machines, automated fibre placement (AFP) machines, and automated tape laying (ATL) machines. Further details follow for each of these AMP techniques.

1.12.1 Filament winding (FW)

FW machines essentially wind filament (tow or roving) sequentially in tension onto a rotating mandrel. The benefit of the FW process is in the ease of manufacture of composite hollow structures such as cylinders and pressure vessels. The winding patterns are predetermined and winding ordinarily displays high geometrical precision. The filament tow may be either resin infused during the process of winding (wet machines), or wound as prepreg tows. The mandrel typically has rotational freedom about at least one-axis, often two or more, and filaments are delivered from a synchronised oscillating delivery head. Fibre orientation can be guided via the delivery head, with hoop, helical and polar orientations being most common. Hoop orientations are circumferential (90° to the rotating mandrel), helical orientations are typically ± 45° relative to the rotating mandrel, and polar orientations are longitudinal (0° to the rotating mandrel, i.e. parallel to the mandrel axis). Once a desired thickness is achieved, the FW process is stopped and the resin is cured, typically in an oven or through external heat sources. Axial winding (AW) is a specialised filament winding process that aligns filaments parallel to the mandrel axis.

FW machines can be of either the continuous or discontinuous type. Continuous FW is used to manufacture pipes of varying sizes, though the tendency is to manufacture large diameter pipes. Discontinuous FW contrarily tends to be used to manufacture complex components, pressure vessels and high pressure components. The continuous FW machines are usually two-axis machines unlike discontinuous FW machines which, due to the need to manufacture at greater levels of complexity, are typically multi-axial. Carbon and glass fibre tows are the most commonly used in the FW processes. A benefit of the FW process is that it is high output, since the method is fairly simple. There are drawbacks nevertheless, one example being that concave surfaces cannot be manufactured using FW machines, since the process relies on tensioning the tows while they are laid [19].

1.12.2 Automated fibre placement/advanced fibre placement (AFP)

AFP is a process by which multiple layers of tow (prepreg or slit tape) are laid down in parallel. AFP is a layer-by-layer process and bears similarity to FW techniques since the tool or mandrel may be rotated to assist tow placement. At the same time, the AFP process uses a delivery head to lay the tow, which in turn applies pressure to the tow to maximise its fit to the mandrel. This bears similarity to the process of automated tape laying (ATL), which also uses a loaded deposition head. As a result, this ‘coupled’ layering technology is able to cover complex surfaces and tight curvatures. Prepreg tow or slit tapes are typically smaller in width (3–13 mm) than for ATL machines [19]. AFP machines have the capacity to place 12–32 tows, which can be placed individually. If placed at the same time, AFP machines usually work to a width range, typically between 1.5 and 4 inches. Laid tows are heated and compacted to cure the composite part. While more complex geometries can be manufactured using AFP methods, they cannot achieve the same rates of tow deposition as ATL machines.

1.12.3 Automated tape laying (ATL)

ATL involves the automated laying of wide (3, 6 or 12 inch, or 7.62, 15.24 or 30.48 cm) unidirectional (UD) tapes onto a part mould via a roller system. The process is particularly good for the manufacture of large flat or mildly contoured parts such as wing skins [20]. The roller system is loaded and depending on the complexity of the part, has varying degrees of freedom (up to 10 axes of movement) split between the tooling and the ATL head [20, 21]. ATL is analogous to manual UD tape laying, but is higher precision and completes tap laying jobs in a fraction of the time it would take to lay manually. ATL is a layer-by-layer process, and orientations can be varied (typically 0°, 90°, ±45°). ATL machines are classified as flat, contoured or highly contoured [19], based on the flatness or contour of the tooling surface. An ATL head contains composite spool (usually carbon or glass), a system for tape guidance, a roller (or compaction shoe), and a cutter. Alternating the fibre orientations during the laying of UD tows ensures structural integrity in the final composite cylinder or vessel. ATL is cost effective as it there are no material cutouts (no weaving, stacking, cutting required). The process is concurrently fast as (a) material is laid directly onto a part mould and (b) the use of prepreg circumvents the need for curing preforms that require time-consuming resin injection.

Modern ATL machines benefit from flaw detection sensor technologies, optical positioning, and systems that heat the tape, which in turn improve tack. ATL using diode lasers process speeds are at a rate of several metres per minute. The primary difference between ATL and AFP is that ATL is most efficient when part-curvatures are gentle or non-existent, while AFP is most efficient when parts are highly contoured or tightly curved.

References

[1] Turssi C P, Purqueiro B D M and Serra M C 2003 Wear of dental resin composites: insights into underlying processes and assessment methods–a review *J. Biomed. Mater. Res.* **65B** 280–5

[2] Sannino A P and Rack H J 1995 Dry sliding wear of discontinuous aluminium composites: review and discussion *Wear* **189** 1–19

[3] Patel M, Saurabh K, Prasad V V B and Subrahmanyham J 2012 High temperature C/C-SiC composite by liquid silicon infiltration: a literature review *Bull. Mater. Sci.* **35** 63–73

[4] Cenna A A, Doyle J, Page N W, Beehag A and Dastoor P 2000 Wear mechanisms in polymer matrix composites abraded by bulk solids *Wear* **240** 207–14

[5] Tewari U S, Bijwe J, Mathur J N and Sharma I 1992 Studies on abrasive wear of carbon fibre (short) reinforced polyamide composites *Tribol. Int.* **25** 53–60

[6] Alam P, Robert C and Ó Bradaigh C M 2018 Tidal turbine blade composites—a review on the effects of hygrothermal aging on the properties of CFRP *Composites* B **149** 248–59

[7] Chang F, Lam F and Kadla J F 2013 Application of time–temperature–stress superposition on creep of wood–plastic composites *Mech. Time Depend. Mater.* **17** 427–37

[8] Nishikawa M, Okabe T, Hemmi K and Takeda N 2008 Micromechanical modeling of the microbond test to quantify the interfacial properties of fiber-reinforced composites *Int. J. Solids Struct.* **45** 4098–113

[9] Chen E J H and Young J C 1991 The microdebonding testing system: a method for quantifying adhesion in real composites *Compos. Sci. Technol.* **42** 189–206

[10] DiFrancia C, Ward T C and Claus R O 1996 The single fibre pull out test: 1. Review and interpretation *Composites* **A27** 597–612

[11] Mehdikhani M, Gorbatikh L, Verpoest I and Lomov S V 2019 Voids in fiber-reinforced polymer composites: a review on their formation, characteristics, and effects on mechanical performance *J. Compos. Mater.* **53** 1579–669

[12] Patel N, Rohatgi V and Lee L J 1995 Micro scale flow behavior and void formation mechanism during impregnation through a unidirectional stitched fiberglass mat *Polym. Eng. Sci.* **35** 837–51

[13] Summerscales J 2019 *Composites Design and Manufacture* Plymouth University teaching support materials for MATS347

[14] Dimeski D, Sapaseska D and Srebrenkoska V 2009 Ballistic strength of woven fabrics for personal protection *J. Eng. Process Manag.* **1** 32–45

[15] Hogg P J, Ahmadnia A and Guild F J 1992 The mechanical properties of non-crimped fabric-based composites *Composites* **24** 423–32

[16] Arvinte C, Sandu A V, Burduhos-Nergis D D, Bernevig-Sava M A and Bejinariu C 2019 Technical requirements and materials used in firefighters gloves manufacturing *IOP Conf. Ser. Mater. Sci. Eng* **572** 012070

[17] Eckold G 1994 *Design and Manufacture of Composite Structures* (Woodhead: Cambridge)

[18] Fernandez I, Blas F and Frovel M 2003 Autoclave forming of thermoplastic composite parts *J. Mater. Process. Technol.* **143-44** 266–9

[19] Maccarville D A 2012 Intellectual property, patents and innovation in aeronautics *Innovation in Aeronautics* ed T M Young and M Hirst (Amsterdam: Elsevier) pp 263–304

[20] Campbell C F 2004 *Manufacturing Processes for Advanced Composites* (Amsterdam: Elsevier)

[21] Crosky A, Grant C, Kelly D, Legrand X and Pearce G 2015 Fibre placement processes for composites manufacture *Advances in Composites Manufacturing and Process Design* (Amsterdam: Elsevier) pp 79–92

IOP Publishing

Composites Engineering: An A–Z Guide

Parvez Alam

Chapter 2

‘B’

2.1 Bag moulding (vacuum bag moulding)

Bag moulding, or vacuum bag moulding, is a composites manufacturing process that involves vacuum compression of wet (hand lay-up) laminates or composite prepreg on the surface of a mould. Uniform compression can be achieved as the process is conducted within a bag that intimately contacts the composite surface, subsequently exerting pressure under the influence of the vacuum. The vacuum bag is normally a thin film of polypropylene, polyethylene, PVA, nylon, or mylar (BoPET (biaxially-oriented polyethylene terephthalate)). Bag moulding is typically used to produce structural composites used, e.g. in construction, aerospace, automotive, renewable energy, maritime, and military industries. It is a moulding approach that enables the manufacture of well consolidated, high fibre volume fraction composites with low fractions of air bubble voids.

2.1.1 The process

A prepreg or wet (hand lay-up) laminate is placed across a composite mould. The mould is essentially the same as in an open mould process, and is coated with a release agent and gel lubricant, or a plastic film, depending on the geometrical complexity of the mould. Release film may also be applied to the laminate as well as bleeder material to absorb excess resin from the laminate. Bleeder sheet materials are essentially absorbent sheets with good drapeability, most commonly made from 5 to 10 mm thick permeable fibreglass, non-woven nylon or polyester cloth. Breather material is placed over the bleeder sheet and the edges of the bag are sealed after which a vacuum is applied to the interior of the bagging system, typically at atmospheric pressure. It should be noted that oftentimes, the bleeder and the breather are treated as one and the same, and only one breather/bleeder sheet is

doi:10.1088/978-0-7503-3527-0ch2

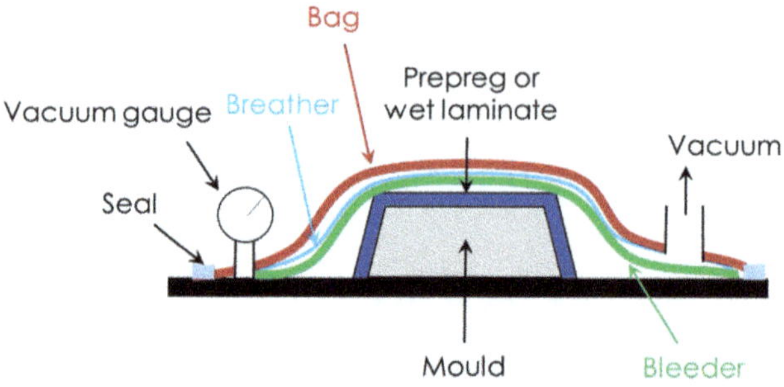

Figure 2.1. Fundamental set up for a vacuum bagging process.

used. The vacuum is held until curing of the composite is complete. Figure 2.1 shows the fundamental set up of a vacuum bagging process.

2.1.2 Benefits of bag moulding

- The process can be used to manufacture large parts of devices, such as wind and tidal turbine blades.
- Atmospheric pressure is applied uniformly across all areas of the material. At this pressure, there is little to no damage of the mould.
- Resin content can be controlled fairly accurately using this process.
- Laminating efficiency is high and, relatively, mould shapes can be easily customised. This is partly due to the precision and close thickness tolerance that can be achieved through bag moulding.
- Bag moulding materials are generally low cost, and the machining and material costs needed for manufacturing moulds are also relatively low.

2.1.3 Disadvantages of bag moulding

- Unlike partly or fully automated processes, the process relies on the operation skills of the user. As such inconsistencies may arise between parts made in different production runs.
- Consolidation is not as good as with compression moulding or autoclaving as the pressures are low (atmospheric pressure is used).
- Many of the materials used in the process are not-reusable and as such the process generates vast quantities of waste.
- The process is relatively slow and may incur significant costs for labour.
- Surface finishing of the composite is usually needed on the bagging side of the material.

Some of the disadvantages, such as slow production times and material waste, can be circumvented by using reusable bag moulding (RBM) techniques. This is a hybrid method that combines aspects from light resin transfer moulding (LRTM) and vacuum infusion processes (VIP). Two variants of this process are used: (1) closed cavity bag moulding (CCBM) and (2) silicone bag moulding (SBM).

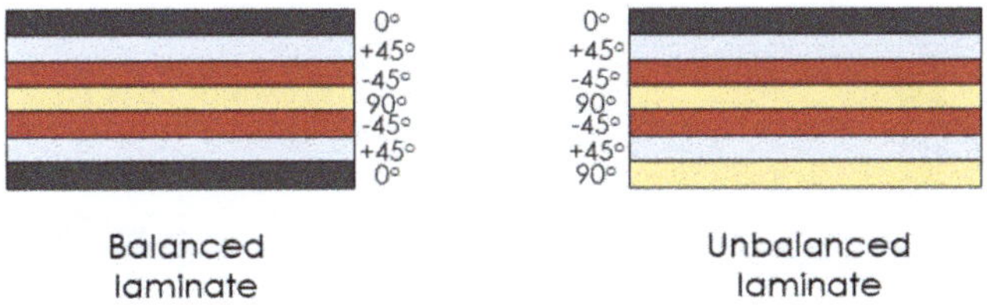

Figure 2.2. Examples of balanced and unbalanced laminated composites.

2.2 Balanced laminates

A balanced laminate is a laminate comprising equal numbers of both $+\theta$ and $-\theta$ plies. For the laminate to be balanced, the 'balancing' plies should additionally be of equal sizes and have the same properties. Balancing laminates helps prevent warping issues during, e.g. cooling processes, since differences in ply angle orientations are known to influence expansion during cure at ply interfaces. Figure 2.2 compares a balanced laminate lay-up against an unbalanced one. The balanced lay-up in figure 2.2 is also a symmetric laminate since the laminates are mirrored about a centre, the laminate being balanced therefore about its centreline and the total number of plies is odd.

In equation (2.1), a balanced laminate would be one where $A_{16} = A_{26} = 0$, and if the balanced laminate is symmetrical with an odd number of plies, then $[B] = 0$. Here, the forces, N, and moments, M, can be calculated from the extensional stiffness (A_{ij}), the coupling stiffness (B_{ij}), the bending stiffness (D_{ij}), the normal and shear strains, ε_x, ε_y, ε_{xy}, and the laminate curvatures, κ_x, κ_y, κ_{xy}, i.e. $\frac{1}{r}$ where r is the radius of curvature.

$$\begin{Bmatrix} N_x \\ N_y \\ N_{xy} \\ M_x \\ M_y \\ M_{xy} \end{Bmatrix} = \begin{bmatrix} A_{11} & A_{12} & A_{16} & B_{11} & B_{12} & B_{16} \\ A_{12} & A_{22} & A_{26} & B_{12} & B_{22} & B_{26} \\ A_{16} & A_{26} & A_{66} & B_{16} & B_{26} & B_{66} \\ B_{11} & B_{12} & B_{16} & D_{11} & D_{12} & D_{16} \\ B_{12} & B_{22} & B_{26} & D_{12} & D_{22} & D_{26} \\ B_{16} & B_{26} & B_{66} & D_{16} & D_{26} & D_{66} \end{bmatrix} \begin{Bmatrix} \varepsilon_x^0 \\ \varepsilon_y^0 \\ \varepsilon_{xy}^0 \\ \kappa_x \\ \kappa_y \\ \kappa_{xy} \end{Bmatrix} \tag{2.1}$$

Balanced laminates should not be confused with **balanced construction**, which relates to a fabric comprising equal quantities of warp and weft.

2.3 Band width

There are several instances in composites engineering where the term *band width* (or bandwidth) is used. This can be a cause for confusion and as such, it may be beneficial to elucidate the terminological differences within composites engineering.

2.3.1 In: filament winding

In filament winding, band width refers to the width of the fibre tow or roving that is laid down over the revolving mandrel. The bandwidth is intimately tied to the

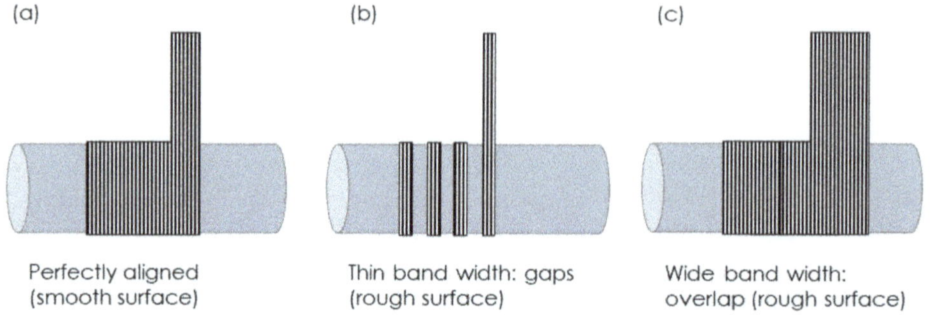

Figure 2.3. Importance of band width in filament winding (a) perfect size bandwidth relative to number of circuits for 'closure' results in a smooth surface with each tow adjacent and flush with the next (b) the same number of circuits for 'closure' using thinner bandwidth tow results in gaps and a final rough surface and (c) the same number of circuits for 'closure' using thicker bandwidth tow results in overlap and a final rough surface.

number of circuits to closure. A perfectly flush and smooth surface is achieved through coupling a correct number of circuits to closure with the bandwidth, figure 2.3(a). Decreasing the bandwidth from this results in gaps between the tows and a rough surface, figure 2.3(b), while increasing the bandwidth results in overlap and a consequent rough surface, figure 2.3(c).

2.3.2 In: piezoelectric composites

In a continuous frequency band, bandwidth defines the difference in frequency between the upper and lower limits of the band. The unit of bandwidth is thence, Hertz (Hz). Piezoelectric materials can (a) produce an electric current when subjected to a mechanical stress and (b) deform (usually to no more than 4% strain), when an electric current is passed through them. This is represented in matrix form in equation (2.2), where **D** is the electric displacement vector, ε is the strain vector, **E** is the applied electric field vector, σ is the stress vector, e^{σ} is the dielectric permittivity, d^{d} and d^{c} are piezoelectric coefficients, and s^{E} is the elastic compliance.

$$\begin{Bmatrix} \mathbf{D} \\ \varepsilon \end{Bmatrix} = \begin{bmatrix} e^{\sigma} & d^{d} \\ d^{c} & s^{E} \end{bmatrix} \begin{Bmatrix} E \\ \sigma \end{Bmatrix} \tag{2.2}$$

Examples of 'naturally occurring' piezoelectric materials include Berlinite, cane sugars, Rochelle salts, quartz, silk, tendon, enamel, dentine and apatite crystals in dry bone. Synthetic piezoelectric materials include zinc oxide, PZT (lead–zirconate–titanate), lead titanate, sodium tungstate, and barium titanate. Piezoelectric composites are piezoelectric materials embedded into a matrix, which broadly range from thermoplastic polymers to cementitious matrices.

Since piezoelectric materials effectively convert mechanical energy to electrical energy, and vice versa, they are often referred to as piezoelectric transducers. It is important to be able to estimate the rate at which a piezoelectric transducer is able to change length under an applied electric current, especially in high speed applications. Bandwidth can be estimated if at least (a) the maximum current that can be produced is known (b) the load capacitance of the piezo is known (c) the desired signal amplitude is known.

2.4 Batt

Batt is a (textile) fabric or structure that has been built by compressing fibres. The process entails the compression of material fibres to create an interlocked network of fibres that is able to retain its structure. There is no spinning or weaving involved in the manufacture of a batt and as such it has the advantage of manufacturing simplicity. Fibre orientation can be controlled to some extent in a batt and both directional and random fibre arrangements are commonly manufactured. The fibres in a batt are entangled by one of three means.

1. **Mechanical bonding**: fibres in non-woven batts can be mechanically entangled through friction-based processes. These processes enable the strengthening of the fibre network in a batt and may include needle punching, hydro-entangling, or stitch-bonding. In needle punching, the fibre web is moved continuously through an array of needles that punch through the web to forcibly entangle the fibres. Parameters that affect the quality of the batt include the depth of penetration and the punch density (number of needles per unit area). Hydro-entangling entails the use of thin high-pressure water jets to forcibly entangle the fibre web (known also as spunlacing). Compared to other non-wovens, hydro-entangling results in good drape, strength and handle. Stitch-bonding involves the stitching of the fibre web using a continuous filament. The filament is stitched into loops that hold the fibre web together [8].
2. **Thermal bonding**: widely used for low melting-point thermoplastic fibres. The method essentially interlocks fibres through heat activated adhesion. During thermal bonding, the batt is passed through heated metallic cylinders and bonding occurs between melted cross over points, and this is further encouraged by extra pressure from movement between the cylinders. Though the process is commonly used for, e.g. polypropylene and polyethylene thermoplastic fibres, powdered binder can also be added to a batt to adhere, e.g. natural fibres at cross over points.
3. **Chemical bonding**: typically involves the application of a liquid bonding agent to fibres in the fabric. The bonding agent can be impregnated into the fibre web, spray-coated onto the fibres, print-bonded, or applied to the fibre web via a foam.

There are benefits and disadvantages to batts and these are listed in table 2.1.

Batts are typically laminated to one another, either through parallel laying (where all batts are laid in the machine direction), or through cross laying (where batts are laid as $[0°/90°_n]$), i.e. machine direction and cross direction lay-ups.

Table 2.1. List of general benefits and disadvantages of batts.

Benefits	Disadvantages
Easy manufacture	Low mechanical properties
High productions rates	Low longevity
Can be moulded to surfaces easily	High equipment costs
Drapeability	
Fibres can be directional or randomly oriented	
Can use recycled thermoplastics	
Low cost	
Many areas of application	

2.5 Biaxial Load

Uniaxial testing is a measurement method that allows the user to develop a fundamental understanding of composite properties and mechanical behaviour. Many applied engineering composites are nevertheless, routinely subjected to loading in more than one direction. Examples can be as varied as composite bridge reinforcements, composite boat hulls, and composite monocoques. In such instances, uniaxial tests cannot capture realistic mechanical behaviour and alternative testing methods may be used preferentially, to develop a more detailed understanding on composite properties and mechanical behaviour. Biaxial testing is one such method that can help develop greater insight into the mechanical behaviour of the structure as it would be loaded in service-conditions, especially with materials such as fibre reinforced plastics, which may exhibit a range of properties from orthotropy to quasi-isotropy. The method involves the axial loading of a flat specimen in two orthogonal directions. The shape of the specimen can vary from rectilinear plates to more commonly used cruciform plates. The variability of specimen shape reflects the fact that there are no available standards for the biaxial testing of composites from any of the international standardisation organisations. Though biaxial testing elucidates more information on the properties and behaviour of composites, there is still a drawback, in that many composites loaded under service-conditions are in fact not flat and though an improved understanding of a composite material can be obtained through biaxial testing, the transition from laboratory to service-conditions are different and failure not predicted can commonly occur at corners, curves, bends and when there may be extra-loadings such as torsional loads and bends. Figure 2.4 shows (a) a biaxial cruciform tensile test set up and (b) a schematic of a cruciform specimen [9]. Biaxial/multiaxial tests can also be conducted on edge clamped thin elliptical or round plates, figure 2.5, and thin tubular specimens, figure 2.6. The elliptical/round plates are essentially pressurised (like a burst test) and multiaxial stresses result from pressurisation. Tubular specimens can be subjected to both axial loading and torsion simultaneously, or axial loading and internal pressure loading (compression).

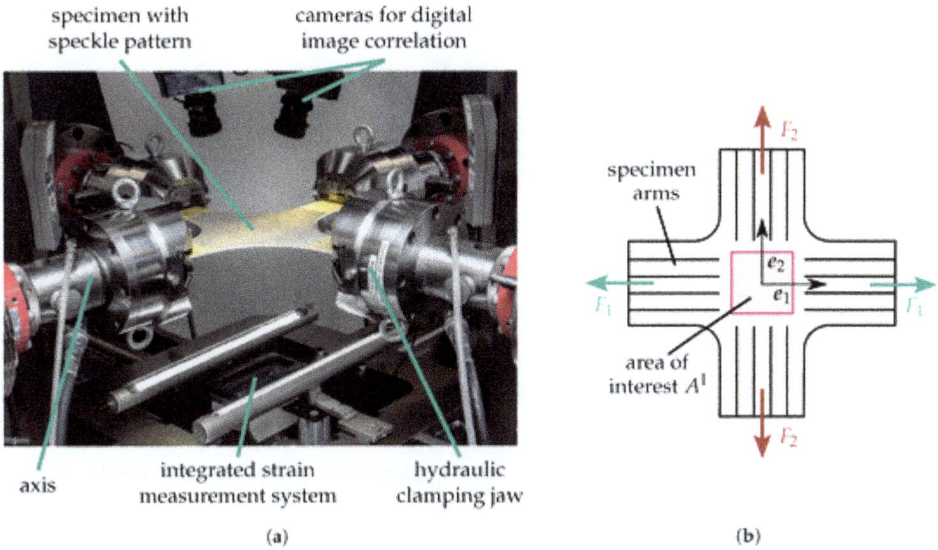

Figure 2.4. Biaxial tensile testing set up and an exemplary cruciform specimen. (a) Biaxial tensile testing device; (b) Cruciform specimen. Reprinted from [9] with permission from MDPI (CC BY 4.0).

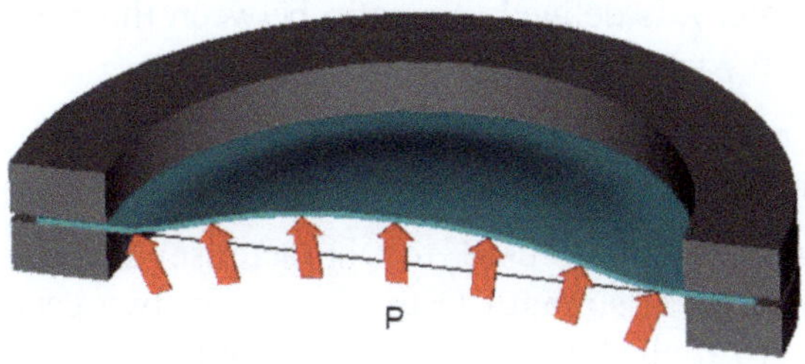

Figure 2.5. Round or elliptical flat sheet used in a biaxial bulge test. Reproduced from [10] InTech Open (CC BY 4.0).

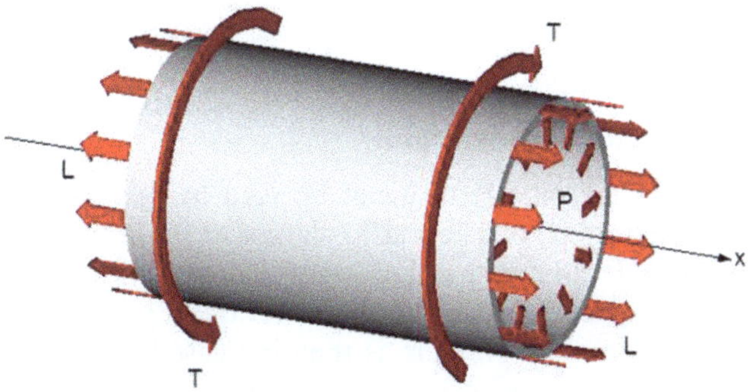

Figure 2.6. Thin tubular specimen with combined axial-torsion loading, or axial-compression (from internal pressure) loading. Reproduced from [10] InTech Open (CC BY 4.0).

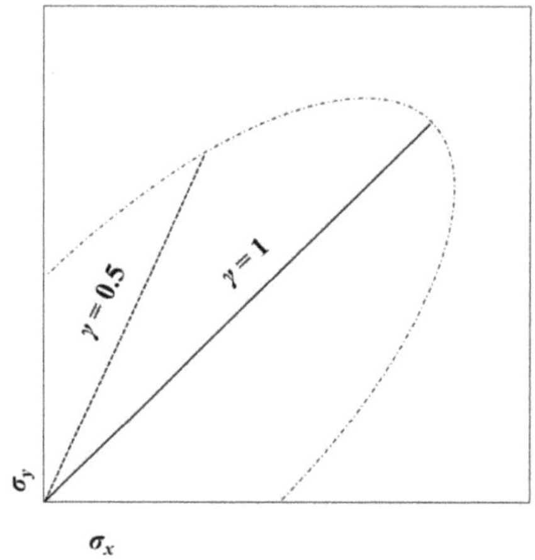

Figure 2.7. Schematic example of a typical failure stress envelope from a biaxial tensile test (γ shown by lines for 0.5 ($\sigma_y > \sigma_x$) and 1 ($\sigma_y = \sigma_x$)).

Biaxial testing can be beneficial for the analysis of composites with a minimum of two fibre orientation axes. Stresses are typically plotted in two orthogonal axes (e.g. x and y) yielding a stress envelope as shown in the example schematic of figure 2.7. The biaxial stress state, γ, is defined as a ratio between these orthogonal (principle) stress states such that $\gamma = \frac{\sigma x}{\sigma y}$.

2.6 Burst strength

Burst strength testing is often used to characterise the pressure resistance of composite structures. Burst strength testing comes in two general forms The first is a hydraulic pressure applied internally within a system until rupture. This is often used for filament-wound composites structures with constant thickness such as pipes and vessels. The second can be used to test flat samples of woven composites or fabrics. The two common tests in use are the Mullen test and the ball burst test. The following subsections will cover each of hydraulic pressure testing, the Mullen test and the ball burst test.

2.6.1 Hydraulic testing for burst strength

A common test set up for pipe testing can be seen in figure 2.8. A steel shaft is inserted into the composite pipe. The shaft is larger at its ends and a hydraulic seal is applied to these in order to create a fluid-tight seal with the composite pipe. Oil is then pumped into the space between the shaft and the pipe at pressure through an inlet inside the steel shaft. The pressure is then increased until the composite pipe ruptures, at its burst strength. Useful standards include: the ASTM D1599-18 [1] and the ASTM D1598-15a [2], and ISO 14692-1:2002 [3]. For closed-end composite pressure vessels, pressurisation is achieved by pumping oil directly into the nozzle of the vessel until rupture.

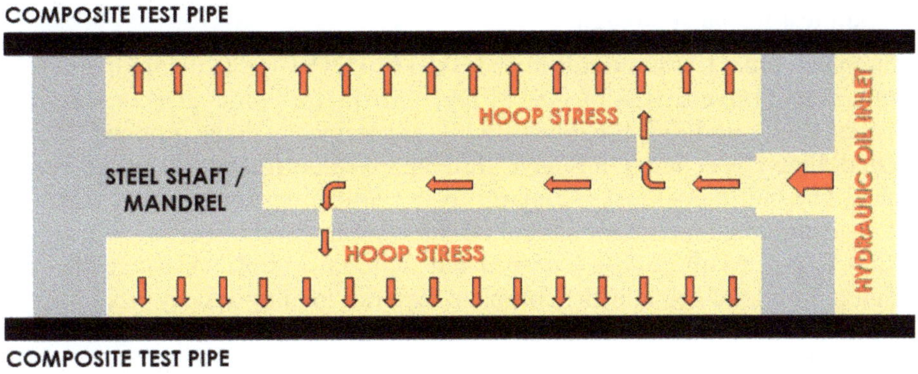

Figure 2.8. Fundamental set up for a hydraulic pipe test.

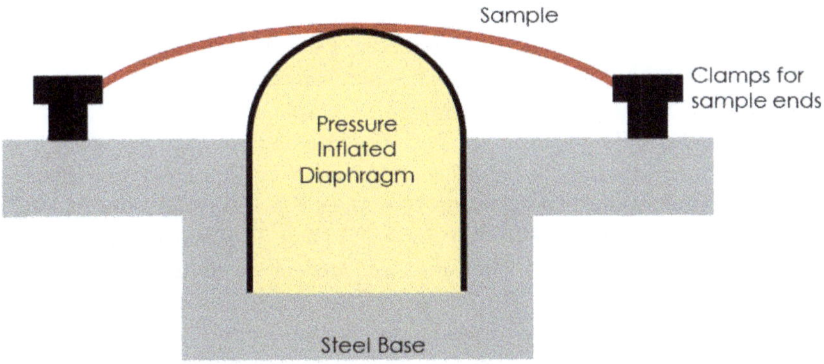

Figure 2.9. Fundamental set up for a Mullen test.

2.6.2 Mullen test for burst strength

Burst strength can be determined in wovens and textiles using a Mullen test. The test is essentially a measurement of material resistance to rupture (or bursting) from pressure applied by a hydraulic or pneumatic diaphragm. The test method is versatile and a diverse range of materials can be tested, including wovens. The Mullen test system includes the inflatable diaphragm for pressure application (usually made of synthetic or natural rubber), clamps to hold the specimen in place, a pressure gauge and pressure system (usually via glycerine or ethylene glycol as a substitute, or if pneumatic, then air), and pressure recording instruments that provide the pressure at the instant of rupture. Figure 2.9 provides a schematic to highlight the Mullen test set up.

ASTM D3786/D3786M-18 [4] is the relevant test standard for Mullen testing of textiles/fabrics. There are nevertheless several other Mullen test standards prepared for different material sheets including: ASTM D774-97, TAPPI T403 om-97, and ISO 2758 for paper; ASTM D751-00, ASTM D751-00 and ASTM D3393-91 for coated fabrics; ASTM D3656-97 for louver cloth woven from vinyl coated glass fibre yarn, and ASTM D2210-00 for leather.

2.6.3 Ball test for burst strength

Common ball burst test standards include ASTM D3787-16(2020) [5] for textiles at a constant rate of traverse (CRT), ASTM D6797-15 [6] for fabrics at a constant rate of extension (CRE), and ISO 12625-9:2015 [7] for specifically tissue products. It should be noted that CRT and CRE give different results against the same material samples. The ball burst works similarly to the Mullen test except that rather than using an expanding diaphragm under hydraulic or pneumatic pressure, it uses a steel ball that is pushed through a clamped flat material sample. As such, the ball is attachable to a load cell and the complete material loading history recorded.

References

[1] ASTM D1599-18 2018 *Standard Test Method for Resistance to Short-Time Hydraulic Pressure of Plastic Pipe, Tubing, and Fittings* (West Conshohocken, PA: ASTM Int.)

[2] ASTM D1598-15a 2015 *Standard Test Method for Time-to-Failure of Plastic Pipe Under Constant Internal Pressure* (West Conshohocken, PA: ASTM Int.)

[3] ISO 14692-1:2002(en) 2002 *Petroleum and Natural Gas Industries—Glass-reinforced Plastics (GRP) Piping—Part 1: Vocabulary, Symbols, Applications and Materials*

[4] ASTM D3786/D3786M-18 2018 *Standard Test Method for Bursting Strength of Textile Fabrics–Diaphragm Bursting Strength Tester Method* (West Conshohocken, PA: ASTM Int.)

[5] ASTM D3787-16(2020) 2020 *Standard Test Method for Bursting Strength of Textiles—Constant-Rate-of-Traverse (CRT) Ball Burst Test* (West Conshohocken, PA: ASTM Int.)

[6] ASTM D6797-15 (2015) 2015 *Standard Test Method for Bursting Strength of Fabrics Constant-Rate-of-Extension (CRE) Ball Burst Test* (West Conshohocken, PA: ASTM Int.)

[7] ISO 12625-9:2015 2015 *Tissue paper and tissue products—Part 9: Determination of ball burst strength*

[8] Kane F 2009 Non-wovens in smart clothes and wearable technologies Smart Clothes and Wearable *Smart Clothes and Wearable Technology* ed J McCann and D Bryson (Woodhead: Cambridge)

[9] Schemmann M, Lang J, Helfrich A, Seelig T and Bohlke T 2018 Cruciform specimen design for biaxial tensile testing of SMC *J. Compos. Sci.* **2** 12

[10] Escarpita D A A, Cardenas D, Elizalde H, Ramirez R and Probst O 2012 Biaxial tensile strength characterisation of textile composite materials *Composites and Their Properties* ed N Hu (Rijeka: InTech)

IOP Publishing

Composites Engineering: An A–Z Guide

Parvez Alam

Chapter 3

‘C’

3.1 C-scan

Non-destructive testing (NDT) comprises a broad group of analysis techniques, many of which are used to valuate the properties of composite materials without damaging it. C-scans are one such NDT technique commonly used to evaluate the health of composites by identifying defective regions, e.g. air-bubbles, de-bonded areas, fibre misorientation, fracture paths, and other defects. C-scanning provides information on the sizes and locations of defective regions in a composites. This technique is typically used as a quality inspection method, but is also used as a detection tool for low technology readiness level (TRL) R&D.

C-scanning is an ultrasonic NDT technique, where piezoelectric transducers convert an electrical voltage into a mechanical vibration, consequently generating ultrasonic waves (in the MHz range). Two types of ultrasonic wave propagate through a composite: longitudinal waves (compression) and transverse waves (shear). Longitudinal waves travel almost 2-fold faster that transverse waves. The velocity at which waves travel through an elastic material are a function of material density and stiffness. Waves travelling through anisotropic and orthotropic composites (e.g. FRPs) will therefore experience different wave velocities as a function of the scanning direction.

Unlike homogeneous, isotropic materials, composites contain interfaces. As such, both reflection and transmission of the ultrasonic wave can occur at material boundaries in composites, examples including: fibre–matrix interfaces and interfaces between laminae. The ratio of transmitted energy to reflected energy will depend on the material density, the velocity at which the wave is travelling, and the angle of incidence of the ultrasonic wave. Assuming the angle of incidence is normal to a surface, reflection and transmission coefficients, α_r and α_t respectively, can be represented by the following equations: $\alpha_r = \frac{I_r}{I_i} = \frac{(z_2 - z_1)^2}{(z_2 + z_1)^2}$ and $\alpha_t = \frac{I_t}{I_i} = \frac{4z_2z_1}{(z_2 + z_1)^2}$, where I_r, I_t and I_i are the intensities of the reflected, transmitted and incident waves,

doi:10.1088/978-0-7503-3527-0ch3

respectively, and z_1 and z_2 are the acoustic impedance of materials 1 and 2 of the composite, respectively. Acoustic impedance of a material is a product of the wave velocity and material density. While impedance is fairly simple to determine in a homogenous, isotropic material, it is considerably more complex in composite materials where both the density and wave velocity vary randomly through the structure and the differences can be large. Using glass fibre reinforced epoxy to highlight the differences, while for epoxy resin $z \approx 3 \times 10^5$ g cm^{-2} s^{-1}, for glass $z \approx 15 \times 10^5$ g cm^{-2} s^{-1}. Additionally, while for epoxy resin $\alpha_r = 0.045$ and $\alpha_t = 0.955$, for glass $\alpha_r = 0.248$ and $\alpha_t = 0.752$.

An additional challenge with C-scanning in composites are the losses that arise as a function of internal wave scattering and reflection at interfaces, and absorption in bulk matter. This is termed attenuation, and high attenuation means that C-scanning composites cannot easily be conducted over long distances (or through very thick samples). The attenuation transmitted through a thin composite sample can be monitored using a through-transmission method to enable the detection of defective regions in the composite. In this method, sonic waves are emitted by a transducer on one side of the composite sample, and received on the other side by a receiver. Since wave transmission can be monitored (and attenuation can be deduced), high levels of attenuation using this method are indicators to the presence of defective regions in the composite.

3.2 Carbon fibre

Carbon fibres are manufactured for different applications and as such, carbon fibres on the market exhibit a broad variety of properties. The properties of the fibres are a result of both their process of manufacture, and their raw material source. There are three main methods by which means carbon fibres are classified. The first is based on their properties, the second is based on the precursor fibre material, and the third is based on the final stage of carbon fibre heat treatment. Precursor carbon fibre classifications include those that are: PAN (polyacrylonitrile) based, pitch based (oil or coal), and rayon based. Over 90% of the world's commercial fibres are PAN based with the common diametral range being between 5 and 7 μm.

The typical PAN fibre manufacturing process begins with PAN polymerisation. This is followed by wet spinning, stabilisation, carbonisation and graphitisation. Pitch fibres begin with pitch preparation, and this is followed by melt spinning, infusibilisation, carbonisation and graphitisation. Reactive compounds can then be used to activate the fibre surfaces. Sizing (usually epoxide based) is used to coat commercial fibres. The sizing creates topographical profile (roughness) on the fibre surface and improves the adhesive strength of carbon fibres with polymers.

The tensile modulus, E_t, and tensile strength, σ_t, of carbon fibres are a useful means to their classification. The main classifications are low modulus carbon fibres (LM), standard modulus carbon fibres (HT), intermediate modulus carbon fibres (IM), high modulus carbon fibres (HM), and ultra high modulus carbon fibres (UHM). The tensile modulus values and tensile strength ranges for each follow, and are plotted as an Ashby chart in figure 3.1. LM (E_t 40–200 GPa; σ_t 1–3.5 GPa),

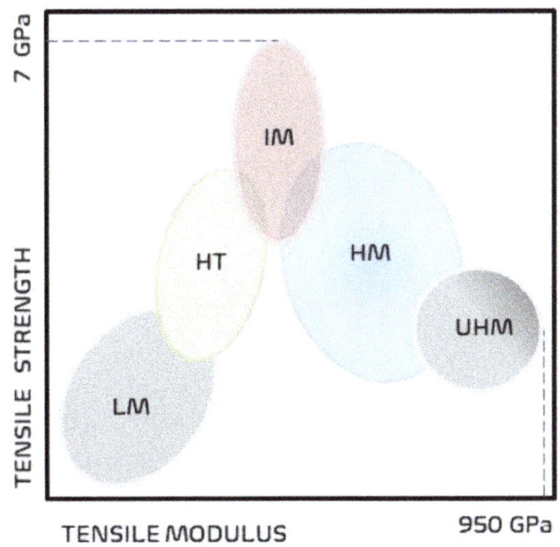

Figure 3.1. Ashby plot showing strength and modulus ranges for different types of carbon fibre.

HT (E_t 200–300 GPa; σ_t 2.5–5 GPa), IM (E_t 280–350 GPa; σ_t 3.5–7 GPa), HM (E_t 350–600 GPa; σ_t 2.5–5 GPa), and UHM (E_t 600–950 GPa; $\sigma_t \approx 4$ GPa).

In terms of heat treatment, carbon fibres are classified into Type I, II and III. Type-I fibres involve high heat treatments, HTT, (>2000° C) and are typical treatments for HM and UHM fibres. Intermediate heat treatments, IHT, (≈1500° C), are termed Type-II fibres and are typically applied to IM and HT fibres. Type-III fibres are subject to a low temperature heat treatment, LHT, (<1000° C) and are typical treatments for the LM fibres. As a result of the lower temperatures used in type-III fibres, they end up being low in terms of cost and are manufactured in high volume.

Carbon fibres end up crystalline, amorphous or partly crystalline as a result of the different heat treatments. Generally speaking, higher heat treatments induce greater levels of crystallinity in the fibres. Crystal orientation and the fraction of crystallinity are intrinsically connected parameters that influence the final properties of the fibre. Crystalline regions in both UHM and HM carbon fibres are oriented in the axis of the fibre direction, giving rise to both high modulus and strength. IM and HT fibres have a large fraction of crystalline regions oriented parallel to the fibre axis, and LM crystalline regions in LM fibres are generally oriented randomly, resulting in lower modulus and strength properties of this fibre type. Table 3.1 provides details on the properties of some commercial carbon fibres [1].

3.2.1 Carbon fibre reinforced plastic (CFRP)

Carbon fibre reinforced plastic (CFRP) is an advanced composite material with specific stiffness and specific strength values that are considerably higher than those of engineering metals such as steel alloys and aluminium alloys. Carbon fibres (such as PAN based fibres) are available in yarns that are wound onto bobbins following chemical, mechanical and thermal treatments, and which can then be embedded into

Table 3.1. Mechanical properties of a range of commercial carbon fibres. Reprinted from [1] with permission from Elsevier. Copyright (2017) Woodhead Publishing.

Sample ($g\ cm^{-3}$)	Density (GPa)	Tensile strength (GPa)	Modulus (%)	Elongation
T300	1.76	3.53	230	1.5
T300J	1.78	4.21	230	1.8
T400H	1.8	4.41	250	1.8
T700S	1.8	4.9	230	2.1
T1000G	1.8	6.37	294	2.2
M35J	1.76	4.7	343	1.1
M40J	1.77	4.41	377	1.2
M46J	1.84	4.21	436	1
M50J	1.88	4.12	475	0.8
M55J	1.91	4.02	540	0.8
M60J	1.94	3.92	294	1.3
M30	1.7	3.92	294	1.3
M30S	1.73	5.49	294	1.9
M30G	1.73	5.1	294	1.7
M40	1.81	2.74	392	0.7
M50	1.91	2.45	490	0.5

a polymer matrix. Fibres are also made available as woven materials, which can be used as a single reinforcing in a polymer matrix. Carbon fibres are also commonly pre-impregnated with uncured polymer matrix, which are typically stored at sub-zero temperatures (to reduce any chances of curing) and can be cut to shape, oriented and laminated as needed. Prepreg materials are typically moulded and cured under heat and pressure to achieve a desired laminated composite.

The aerospace industry is currently the largest consumer of CFRP ($\approx$30%), with wind and sports industries following at about 14% each of the global demand [2]. Both the Boeing 787 Dreamliner and the Airbus A350XWB are made up of $\approx$50% composite material (both GFRP and CFRP), with the remainder of the rest being made up of steel, titanium and aluminium alloys. The 787 Dreamliner comprises $\approx$ 32000kg of CFRP [3], most prominently in the fuselage, the wings, the engineer nocelle/pylon and the horizontal stabilisers and the fin and rudder. CFRPs are also extensively used in the automotive industries (e.g. racing car monocoques), in construction for the strengthening and repair of concrete, steel and timber constructions, and in the sporting equipment industries where CFRP is routinely used for the manufacture of fishing rods, surfboards, racing bike frames, lightweight tennis and badminton racquets, and others.

3.3 Cellular solids

Cellular solids are comprised of assemblies of cells, which by definition are structures with solid edges or faces, containing an enclosed space of air or another material.

There is a plethora of examples of cellular solid in nature including wood, bone, coral, sponges, and cork. Engineering cellular solids are, for the most part, variations of either foams or honeycomb structures. Due to the methods by which they are manufactured, foams are essentially poly-disperse packed cells, while honeycomb structures are systematically packed cells. Both honeycombs and foams are commonly sandwiched between parallel laminates and are used to create lightweight composite structures that work well under specific mechanical and dynamic loads such as bending and impact. The following sections will provide further details on both honeycomb structures and foams.

3.3.1 Honeycomb structures

Honeycomb structures are manufactured in a variety of common arrays, figure 3.2. These include hexagonal, cell-centre reinforced hexagonal, over-expanded, under-expanded, square-cell, and flex-core. The most typical arrangement of a honeycomb core sandwich panel is such that the out-of-plane direction of the honeycomb is perpendicular to the axis of the panels, which are adhered to the open surfaces of the honeycomb cells. The honeycomb core thence carries the normal and shear loads using its surfaces that are perpendicular to the axis of the hexagonal cells. The elastic modulus of honeycomb core in its out-of-plane direction, E_z, is proportional to the volume fraction, V_f of core material with elastic modulus, E_{core}; such that $E_z = V_f E_{\text{core}}$.

Beam theory can be used to predict the modulus of honeycombs in their less-stiff, orthogonal in-plane directions. Loading the in-plane direction results in considerably higher levels of honeycomb core deformation than in the out-of-plane direction, and far lower levels of loading are possible. The orthogonal elastic moduli of in-plane honeycombs can be most simply represented as E_x and E_y. Assuming loading is in the $y-$ direction of the honeycomb, $E_x = E_{\text{core}}(\frac{t}{l})^3 \frac{\cos\theta}{(1 + \sin\theta)\sin^2\theta}$ and

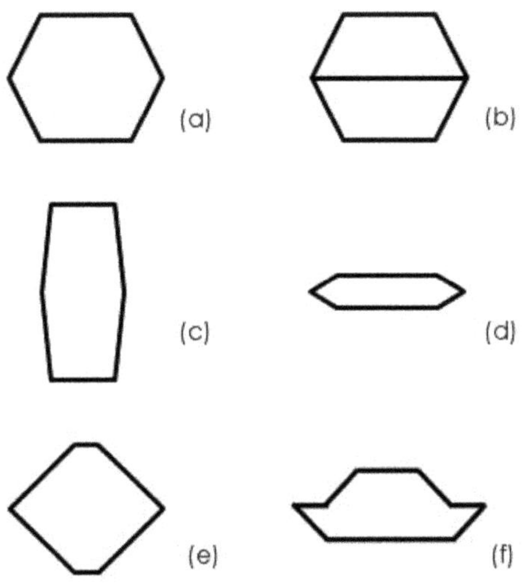

Figure 3.2. Examples of engineered honeycomb structures (a) hexagonal, (b) cell-centre reinforced hexagonal, (c) over-expanded, (d) under-expanded, (e) square-cell and (e) flex-core.

$E_y = E_{\text{core}}(\frac{t}{l})^3\frac{(1+\sin\theta)}{\cos^3\theta}$, where t and l are the thickness and length of the honeycomb cell walls, respectively. The shear modulus, G_{xy}, is related to the elastic modulus of the core material and can be calculated using beam theory as $G_{xy} = E_{\text{core}}(\frac{t}{l})^3\frac{(1+\sin\theta)}{3\cos\theta}$.

3.3.2 Foams

Foam core materials are applied as sandwich core structures in a similar manner to honeycomb cores. One of the main advantages of using foam core as compared to honeycomb core is that foam core panels have a higher capacity for energy absorption under ballistic impact. Foams are classified as either open cell or closed cell. Open cell foams consist of cells that are not entirely encapsulated by material. Closed cell foams consist of cells that are fully encapsulated by material. Open cell foams tend to be softer and more flexible and there is greater capacity for elastic deformation. The reduced density, ρ_{rel}, is an important parameter when predicting the elastic modulus of both open and closed cell foams, and is calculated as a ratio of the effective density, ρ_{eff} and the density of the pure solid material, ρ_s such that $\rho_{\text{rel}} = \frac{\rho_{\text{eff}}}{\rho_s}$. In similitude to honeycomb structures, beam theory can be used to predict the elastic modulus of foams under small deformations. The elastic modulus of foam can be generalised as, $E_{\text{foam}} = E_s C\rho_{\text{rel}}^n$, where E_s is the elastic modulus of the solid material, and C and n are constants dependent on the microstructure of the solid material. It has been empirically understood that generally, in closed cell foams $1 < n < 2$. C is related to periodicity/disorder, such that in cubic arrays $C \approx 1$. Though C and n are essential constants for the prediction of foam elastic modulus, their relationship with microstructure is not well understood. In open cell foams, a common approximation is $C = 1$ and $n = 2$.

3.4 Class (composite class)

Composites are one of four major material taxonomical classes, the others being polymers, glasses/ceramics, and metals, see figure 3.3. Composites can be subdivided into sub-classes including those that are: fibre reinforced, structural, metal matrix, particle reinforced porous, ceramic matrix and polymer matrix. There is inevitable overlap between these sub-classes however, it can still be useful to be able to identify the types of composites that exist within each of the sub-classes. Table 3.2 provides some examples of composite sub-classes and their types.

3.5 Compression moulding

Compression moulding is a process that uses pressure to enable the rapid production of high quality composite parts and can be conducted using 'hot' or 'cold' pressure techniques. Cold pressure techniques are typically used for materials like wood laminates of skateboards decks, where there is no requirement for thermal curing and adhesive curing can be completed at room temperature. Hot compression moulding is more typical of polymer matrix composites where thermal curing is needed.

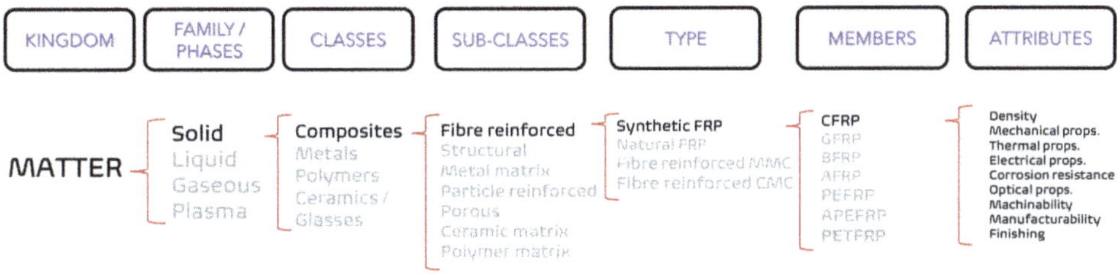

Figure 3.3. Materials taxonomy with a focus on composite materials.

Compression moulds match male and female dies made of metal. Typically, sheet material is placed between the male and female die after which the dies are closed causing the sheet material to take the shape of the mould. In the hot moulding process, the dies are heated and this heat is transferred to the sheet material through the dies. Thermosetting matrix composites are usually loaded in a cold state and cured between the heated dies. The curing cycle, if long and/or complex can delay the product rate of thermosetting matrix composites, and cooling the dies for follow-on moulding further decreases production. Thermoplastic matrix composites are normally loaded in pre-heated dies where they consolidate. Pre-heated dies do not require any cooling and as such result in heightened output (figure 3.4).

Composite parts manufactured by compression moulding typically have high quality surfaces and are usually well consolidated. The process does nevertheless have limitations in that it more favourable to use constant thickness sheet material, pressure can be applied in only one direction, and inward angles cannot be moulded (i.e. part corner angles below 90°).

3.6 Consolidation

During any pressure moulding process, air is squeezed out of the composite, which in turn alters the internal composite microstructure. This is a boundary pressure driven mechanism that leads to a phenomenon called consolidation. Consolidation during the moulding process is critical for the development of high quality composites with optimal mechanical properties. Badly consolidated composites are usually littered with voids and residual stresses, which results in structural discontinuities, weakness and in some cases, warping. The consolidation process is a combination of resin flow mechanisms and fibre deformation/movement. Resin flow and fibre motion is in turn affected by heat, which affects the viscosity of the resin during curing reactions. The consolidation process can to some extent be controlled through the choice of equipment (e.g. autoclave), the applied pressure, the temperature used, and the time over which both pressure and heat are applied.

Resin flow, q, through a preform is often predicted by Darcy's law, equation (3.1), where K is permeability through the preform, μ is the resin viscosity, and $\frac{dp}{dx}$ represents the pressure gradient. The permeability itself is represented by the semi-empirical Kozeny–Carmen relationship, equation (3.2), where r_f is the reinforcing fibre radius, V_f is the fibre volume fraction, and k_0 is an empirical constant which is 0.5–0.7 for longitudinal flow, 11 for transverse flow, 2.7 for ±45° cross plies, and 5.5

Table 3.2. Examples of types of composite under each composite sub-class.

Composite sub-class	Type
Fibre reinforced	Synthetic FRP Natural FRP Fibre reinforced MMC Fibre reinforced CMC
Structural	Structural composite lumber (SCL) Structural timber composites (STC) Reinforced concrete Sandwich panels
Metal matrix	Sheet reinforced Continuous fibre reinforced Short fibre or whisker reinforced Particle reinforced
Particle reinforced	Monodisperse Poly-disperse Flake/large particle Aspect ≈ 1 Aspect > 1 Nano
Porous	Honeycomb core Foam-core Cork-core Coatings Skeletal
Ceramic matrix	Fabric/braid Continuous fibre Discontinuous fibre/whiskers Particle reinforced
Polymer matrix	Fabric/braid Continuous fibre Discontinuous fibre/whiskers Particle reinforced Thermoplastic matrix Thermoset matrix

for textile preforms [4]. The resin viscosity, μ, is represented by equation (3.3), where μ_∞ is a constant, U is the activation energy for viscosity, κ is a constant independent of temperature and α is the degree of cure.

$$q = -\frac{K}{\mu}\frac{dp}{dx} \tag{3.1}$$

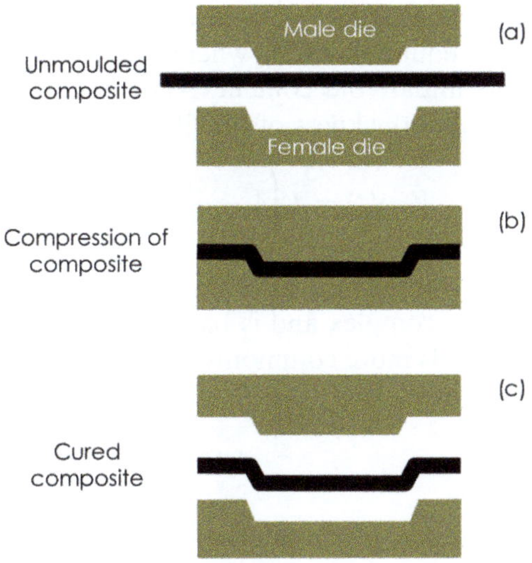

Figure 3.4. Compression moulding (a) pre-moulding, (b) moulding and (c) post-moulding.

$$K = -\frac{r_f^2}{4k_0}\frac{\left(1 - V_f\right)^3}{V_f^2} \tag{3.2}$$

$$\mu = \mu_\infty \exp\left(\frac{U}{RT} + \kappa\alpha\right) \tag{3.3}$$

Consolidation, $-\frac{d(hA)}{dt}$, is essentially a flow equation derived from Darcy's law and is represented by equation (3.4), where K_c is the permeability of the composite and is represented by $K_c = \frac{q_c\mu}{\frac{p_0 - p_c}{h_c}}$, K_b is the permeability of a bleeder sheet and is represented by $K_b = \frac{q_b\mu}{\frac{p_c - p_b}{h_b}}$ (q_b is resin flux through the bleeder, p_0 is the consolidation pressure, p_c and p_b are pressures at the composite and bleeder interfaces, respectively, h_b is the instantaneous thickness of the resin in the bleeder, h_c is the thickness of the composite layers through which resin flux will take place). Further, h_1 is the compacted composite layer thickness and is represented by $h_1 = \frac{h_c}{n}$, h is the total thickness of the composite laminate and n is the number of compacted plies. A is the surface area of the composite, F is the applied force calculated as $\frac{p_0 - p_{atm}}{A}$ (where p_{atm} is the atmospheric pressure), t is time and μ is the resin viscosity.

$$-\frac{d(hA)}{dt} = \frac{\frac{K_c F}{\mu h_1}}{n + \frac{K_c h_b}{K_b h_1}} \tag{3.4}$$

During the process of consolidation, void content is reduced and this can be determined according to equation (3.5), where $V_{\text{void}}(t)$ is the void content with respect to time, V_i is the initial void content, $z(t)$ is the depth of permeation as a function of time and z_c is the thickness of the fibre bed available for flow.

$$V_{\text{void}}(t) = V_i\left(1 - \frac{z(t)}{z_c}\right) \tag{3.5}$$

Fibre movement and compaction also occurs during consolidation however predicting the movement is complex and is based on the fibre network that moves during consolidation. Models more commonly used assume parallel array unidirectional fibres that are in contact.

3.7 Contact moulding

Contact moulding is the oldest thermoplastic matrix moulding processes. It is an open moulding method that results in good surface finishes on the face of the final product in contact with the mould. Figure 3.5 shows the contact moulding process. An open female mould is initially covered in a release agent and gel coat, a dry reinforcing (e.g. chopped strand mat, woven fabrics) is then positioned into the mould after which resin is either poured or sprayed onto it. If poured (hand lay-up) consolidation of the composite is achieved using a mechanical roller, vibroforming (essentially a mechanical roller with a vibrating head), or using gantry robots. Consolidating the preform using a mechanical system helps reduce air bubble voids. If sprayed (spray lay-up), both the fibre and resin are sprayed simultaneously to the

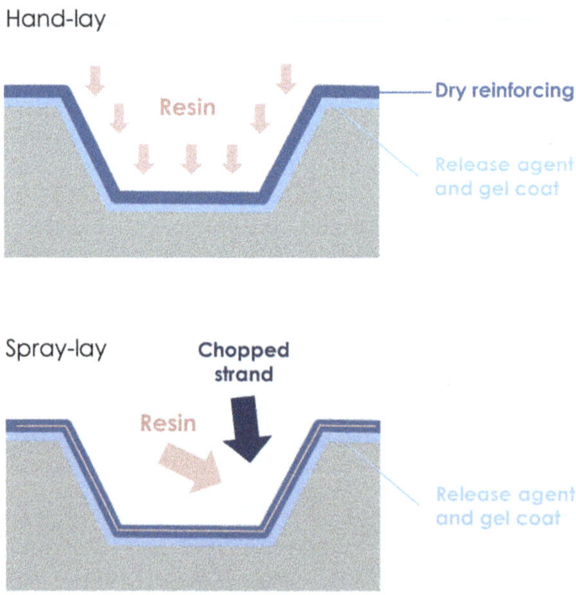

Figure 3.5. Compression moulding (a) pre-moulding, (b) moulding and (c) post-moulding.

inner surface of the mould. Curing takes place at room temperature and atmospheric pressure and an advantage of the process is that further reinforcing (mats/textiles) and resin can be applied to increase the thickness of the product. The product is not size limited and as such very large moulded parts can be manufactured using this process. The process is also energy efficient since neither heat nor pressure are required for curing. The degree of cure of the composite can be assessed using BS2782.

Cripps *et al* [5] list a series of general advantages and disadvantages of contact moulding (hand lay) processes. These are summarised in table 3.3

3.8 Coupon

Composites can be tested at different levels relating to the readiness level of the composite or part. At the higher readiness levels, composite bodies can be tested (e.g. a full scale wind turbine blade) but there are evident limitations to testing at this level since the cost of testing is considerable, and very specialised test equipment is needed. As the readiness level reduces, so too do the financial burdens on testing and at the lowest readiness level (the lab-level), testing is typically carried out using coupons. Since coupons are small and test machines are readily available, coupons can provide valuable insights from a perspective of basic research, that additionally allow for the monitoring of the statistical variability of composite properties.

The type of coupon used will depend upon the specific composite material under scrutiny. The following sections provide examples of common coupons used as a function of type of test and material.

3.8.1 Coupons for tensile testing

Coupons made for tensile testing of composites are amongst the more varied. Fibre reinforced plastics are typically uniform cuboidal coupons with end tabs attached to decrease stress concentrations during clamping and by doing so, minimising the chances of failure due to clamping. Nanocomposites and polymers matrix composites containing micro-fibres or micro-particles are usually tested using tapered, or, dog-bone coupons where the clamped section of the coupon is significantly wider than the central segment of the coupon (the gauge) and similarly to end-tabbed specimens, aims to decrease the chances of failure initiating at the clamp. The BS EN ISO 527 family of standards [6–10] are often used to decide on appropriate coupon specifications for the tensile testing of polymer matrix composites containing reinforcements of different length scales. The ASTM standard ASTM D3039/D3039M-17 [11] is also an oft-used standard for the static tensile testing of polymer matrix composites.

3.8.2 Coupons for compressive testing

In-plane compressive testing of composites can be carried out using ASTM D3410/D3410M-16e1 [12] (a pure shear loading test method). It should be noted that this standard is for unsupported coupon gauge lengths (12–25 mm) and loading is applied at the specimen grips. This is an appropriate test method for polymer matrix

Table 3.3. Advantages and disadvantages of contact moulding (hand lay) processes.

Advantages	Disadvantages
Due to the technique being very old (>60 yrs), the method is already backed by a wealth of knowledge and experience	VOCs are formed (especially from gel coats) and need to be extracted
Standardised training is available	The requirement for manual labour tends to predominate in this process
Short lead times (for finished components)	Operator expertise guide product quality
Tooling costs are minimal	Production rates are low due to low temperature resin curing times
A wide range of suppliers are available for materials and resins	Low viscosity resins are used and are volatile raising health and safety concerns
There is good versatility in the shape and size of the composite	Dimensions are usually inconsistent
Changes to designs can be easily managed	Shrinkage levels are high
Contact moulded parts are cost competitive (short runs)	Possible exotherms in resin-rich regions—typically a bigger problem in thick-section composites
Fibre volume fractions are higher in hand lay than spray lay	Thixotropic agents may be required to drain resin from vertical surfaces
Resin-richness can be an advantage for corrasion resistance	Only one surface is of high quality and the quality is also dependent on the gel coat used
Sandwich panel composites are easily incorporated	Void levels tend to be high as a result of atmospheric pressure curing and manual consolidation methods
	Thermal and mechanical properties are compromised through the use of low molecular weight resins
	Fibre volume fractions are typically very low (0.2 for chopped strand mat, 0.345 for woven fabrics)
	Waste levels are high

composites and coupons are uniform cuboidal and end tabs are beneficial in that they ensure that failure occurs within the specimen gauge length. ASTM D3410/D3410M-16e1 is most appropriate for high polymer matrix composites reinforced by high modulus fibres. A pure end loading method can be used for lower modulus polymer matrix composites containing discontinuous nano/micro particle or fibre reinforcements (ASTM D695 [13]) as these will behave in a manner more similar to rigid plastics. Hydraulic composites test fixtures can also be used in conjunction with ASTM D6641/D6641M-16e1 [14] for balanced and symmetric composites. This is a combined loading method on uniform cuboidal coupons with coupons used where high levels of orthotropy are anticipated.

3.8.3 Coupons for double cantilever beam (DCB) testing

Composites subjected to DCB tests aim to determine the mode I fracture toughness of composite laminates or adhesively bonded composite plates. Coupons for these tests are rectangular span with a single slit present part-way along the span cut across the centre of the specimen. ASTM D5528-13 [15] is an often used standard for FRP specimens and was the first standardisation body to describe DCB test sample dimensions.

3.8.4 Coupons for flexural testing

Coupons for use in three and four point bending tests for use with polymer matrix composites are described in the ASTM D7264/D7264M-21 [16]. The standard includes the use of large span to thickness ratios of 32:1, which is twice the ratio of methods focussed on plastics (ASTM D790 for 3-point bending and ASTM D6272 for 4-point bending).

3.8.5 Coupons for interlaminar shear strength (ILSS) testing

Coupon sizes for interlaminar shear strength testing is advised in the ASTM D2344 standards [17]. This test method is limited to samples with a loading span to thickness ratio of 4:1 where the minimum specimen thickness is 2 mm. Internal stresses that build up in ILSS samples are complex and as such the method cannot entirely be related to interlaminar shear. Nevertheless, the primary failure modes observed in such tests are dominated by resin shear and interlaminar properties, and as such the test is seen as indicative of interlaminar shear properties.

3.9 Crazing

Crazing is a common occurrence in tensile loaded FRP composites and is usually best observed in thermoplastic matrix composites. It is a mechanism within the matrix polymer that allows plastic deformation of the matrix at stress levels that are below the stress levels expected for shear yielding in the matrix. Crazing is essentially the build up of long, thin, hair-like micro-voids in a plane perpendicular to the direction of applied stress, figure 3.6. Hair-like micro-voids tend to form at inhomogeneities in either the polymer, or, at the polymer–matrix interface, and the build up of micro-voids (or crazes) leads to considerable internal air space and the onset of brittle failure. Glassy polymers (or the glassy parts within polymers) are more likely to experience crazing than softer polymers and crazing further reduces the ductility of the matrix polymer. The presence of fibre reinforcements changes the molecular orientation of polymers and in some cases, creates ordered interphases at interfaces. This not only heightens occurrences of inhomogeneities in a composite, but also increases to an extent, rigidity at polymer regions connecting interfaces, which in turn become more prone to crazing. Crazing is not readily observed in thermosetting polymer matrix composites as these are highly cross-linked, which restricts molecular mobility, however, there are examples where crazing has been

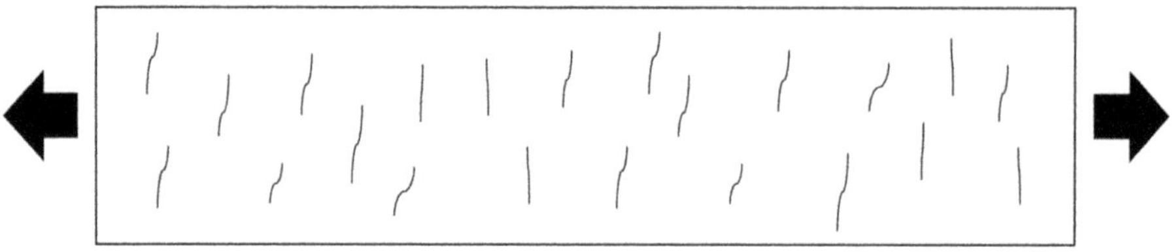

Figure 3.6. Schematic of a tensile coupon with crazing (hair-like micro-voids perpendicular to the direction of loading).

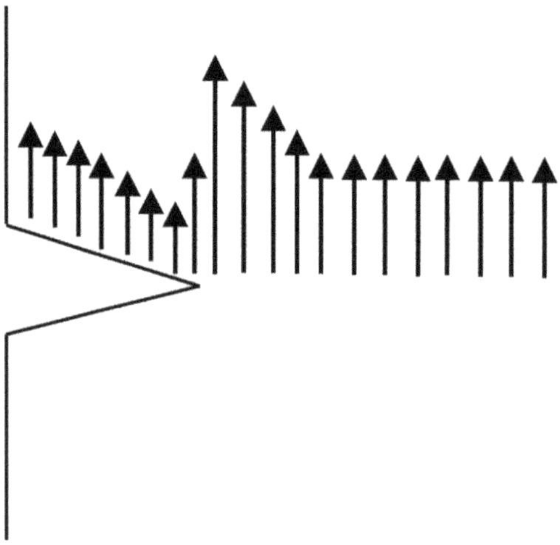

Figure 3.7. Schematic axial stress vectors (black arrows) surrounding a craze.

observed in thermoset–matrix nanocomposites, e.g. CNT reinforced epoxy resin composites [18].

Stress is lower over the surface of a craze than it is in adjacent polymer and there are typically stress spikes at the interface between the tip of a craze and the adjacent polymer. This is illustrated in figure 3.7 schematic where the black arrows indicate axial stress vectors surrounding the craze.

3.10 Crimp

Woven fabrics with a warp (machine direction thread) and a weft (cross direction thread) will exhibit a systematic waviness as a result of weaving. This waviness is referred to as a crimp and its existence in a composite is of importance as failure of polymer matrix composites routinely initiates at crimp points. Both stiffness and strength in composites containing wovens or fabric engineering fibres tend to decrease as a function of increasing crimp. Figure 3.8 provides geometrical details that are helpful in visualising crimp. In this figure n is the number of crimp waves per fibre, l_0 is the side length of one crimp bow, l is the width of one crimp, γ is the wavelength of one crimp and $\gamma = 2l$, α is the angle between the crimp leg and the fibre axis, φ is the crimp angle, L_0 is the extended length of the fibre and $L_0 = 2nl_0$, L is the crimped length of the fibre and $L = nl$, and A is the crimp amplitude [19].

There are a few methods by which means crimp, C_r, can be defined. Crimp can for example, be considered inversely proportional to fibre wavelength as shown in equation (3.6), where L_y is the length of a thread or yarn across its path and γ is its corresponding wavelength. The wavelength can also be used to describe crimp length, c_l, equation (3.7) and crimp frequency, f_c, equation (3.8).

$$C_r = \frac{L_y}{\gamma} \tag{3.6}$$

$$c_l = \frac{1}{2}\gamma \tag{3.7}$$

$$f_c = \frac{2}{\gamma} \tag{3.8}$$

The crimp index, C_i, equation (3.9), which effectively is a ratio of the difference between the straightened fibre, L_0, and the crimped length, L_c, (total length of fibre in fabric) can be expressed as a percentage of the crimped length. A crimp index of 0% indicates there is no crimp, while a crimp index of 100% indicates a complete crimp.

$$C_i = \left(\frac{L_0 - L_c}{L_c}\right) \times 100 \tag{3.9}$$

3.11 Cross ply laminates

In composite laminate theory (CLT), ply stiffness can be assumed to be under a state of plane stress, equation (3.10), where Q_{ij} is termed the reduced stiffness, and σ_1 and σ_2 are normal stresses, σ_6 is shear stress, ε_1 and ε_2 are normal strains, and ε_6 is shear strain. Q_{ij} is related to stiffness, C_{ij} as $Q_{ij} = C_{ij} - \frac{C_{i3}C_{j3}}{C_{33}}(i, j = 1, 2, 6)$. σ for a 0° lamina can be calculated thence as $\sigma_i = Q_{ij}\varepsilon_j$.

$$\begin{Bmatrix} \sigma_1 \\ \sigma_2 \\ \sigma_6 \end{Bmatrix} = \begin{bmatrix} Q_{11} & Q_{12} & 0 \\ Q_{21} & Q_{22} & 0 \\ 0 & 0 & Q_{33} \end{bmatrix} \begin{Bmatrix} \varepsilon_1 \\ \varepsilon_2 \\ \varepsilon_6 \end{Bmatrix} \tag{3.10}$$

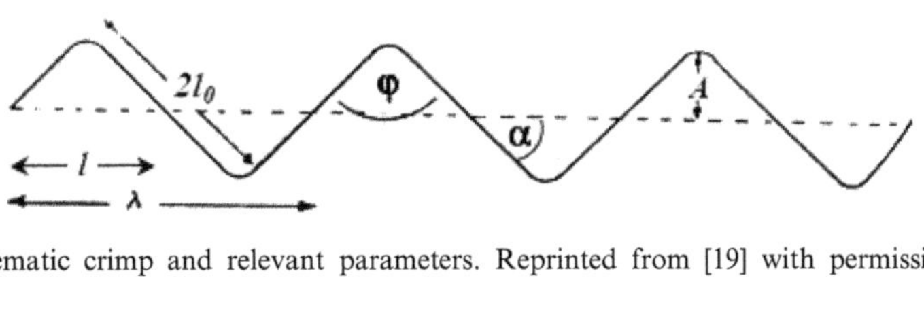

Figure 3.8. Schematic crimp and relevant parameters. Reprinted from [19] with permission of Frontiers (CC BY 4.0).

Q_{ij} relates to engineering constants as: $Q_{11} = \frac{E_{11}}{1-\nu_{12}\nu_{21}}$; $Q_{12} = \nu_{21}Q_{11}$; $Q_{22} = \frac{E_{22}}{1-\nu_{12}\nu_{21}}$; $Q_{16} = 0$; $Q_{26} = 0$; $Q_{66} - G_{12}$, where E is elastic modulus, ν is Poisson's ratio, and G is shear modulus.

Ply angles in laminates are often varied. Calculating the reduced stiffness so as to take into account the ply angle, $\bar{Q}_{ij}$, requires consideration of the effects of these angles on the properties, equation (3.11), where $m = \cos(\theta)$ and $n = \sin(\theta)$, θ being the ply orientation angle.

$$\begin{Bmatrix} \bar{Q}_{11} \\ \bar{Q}_{22} \\ \bar{Q}_{12} \\ \bar{Q}_{66} \\ \bar{Q}_{16} \\ \bar{Q}_{26} \end{Bmatrix} = \begin{bmatrix} m^4 & n^4 & 2m^2n^2 & 4m^2n^2 \\ n^4 & m^4 & 2m^2n^2 & 4m^2n^2 \\ m^2n^2 & m^2n^2 & m^4 + n^4 & -4m^2n^2 \\ m^2n^2 & m^2n^2 & -2m^2n^2 & (m^2 - n^2)^2 \\ mn^3 & -mn^3 & n^3m - m^3n & 2(n^3m - m^3n) \\ mn^3 & -m^3n & m^3n - n^3m & 2(m^3n - n^3m) \end{bmatrix} \begin{Bmatrix} Q_{11} \\ Q_{22} \\ Q_{12} \\ Q_{66} \end{Bmatrix} \tag{3.11}$$

$\bar{Q}_{ij}$ can then be used to determine the extensional stiffness (A_{ij}), equation (3.12), the coupling stiffness (B_{ij}), equation (3.14), and the bending stiffness (D_{ij}), equation (3.13), where t_k is the thickness of the kth lamina and $\bar{z}_k$ is the distance from the mid-plan to the centroid of the kth lamina.

$$A_{ij} = \sum_{k=1}^{n} \left(\bar{Q}_{ij}\right)_k t_k \tag{3.12}$$

$$B_{ij} = \frac{1}{2} \sum_{k=1}^{n} \left(\bar{Q}_{ij}\right)_k t_k \bar{z}_k \tag{3.13}$$

$$D_{ij} = \frac{1}{3} \sum_{k=1}^{n} \left(\bar{Q}_{ij}\right)_k \left(t_k \bar{z}_k^{-2} + \frac{t_k^3}{12} \right) \tag{3.14}$$

These in turn can then be used to calculate the forces, N, and moments, M, according to equation (3.15), which is shown in the expanded form in equation (3.16)

$$\begin{Bmatrix} N \\ M \end{Bmatrix} = \begin{bmatrix} A & B \\ B & D \end{bmatrix} \begin{Bmatrix} \varepsilon^0 \\ \kappa \end{Bmatrix} \tag{3.15}$$

$$\begin{Bmatrix} N_{xx} \\ N_{yy} \\ N_{xy} \\ M_{xx} \\ M_{yy} \\ M_{xy} \end{Bmatrix} = \begin{bmatrix} A_{11} & A_{12} & A_{16} & B_{11} & B_{12} & B_{16} \\ A_{12} & A_{22} & A_{26} & B_{12} & B_{22} & B_{26} \\ A_{16} & A_{26} & A_{66} & B_{16} & B_{26} & B_{66} \\ B_{11} & B_{12} & B_{16} & D_{11} & D_{12} & D_{16} \\ B_{12} & B_{22} & B_{26} & D_{12} & D_{22} & D_{26} \\ B_{16} & B_{26} & B_{66} & D_{16} & D_{26} & D_{66} \end{bmatrix} \begin{Bmatrix} \varepsilon_{xx}^0 \\ \varepsilon_{yy}^0 \\ \varepsilon_{xy}^0 \\ \kappa_{xx} \\ \kappa_{yy} \\ \kappa_{xy} \end{Bmatrix} \tag{3.16}$$

Here, κ_x, κ_y, and κ_{xy} are the laminate curvatures calculated as $\frac{1}{r}$, where r is the radius of curvature. **A cross ply laminate** consists of plies with orthogonal fibre orientations (either 0° or 90°). In these laminates $A_{16} = A_{26} = B_{16} = B_{26} = D_{16} = D_{26} = 0$. Shear and normal forces are thence decoupled in cross ply laminates, and this is also true for both bending and twisting moments. Cross ply laminates can be either symmetric or antisymmetric. If symmetric the $[B] = 0$, but if the laminate is antisymmetric, then $[B]$ is represented by equation (3.17). There is no coupling between the force and moment terms in the case where $[B] = 0$.

$$[B] = \begin{bmatrix} B_{11} & 0 & 0 \\ 0 & -B_{11} & 0 \\ 0 & 0 & 0 \end{bmatrix} \tag{3.17}$$

3.12 Curing

Curing is a process that results in the development of greater cross-linking in thermosetting polymers. In resins used in composites, this reaction is essentially polymerisation (a chemical reaction), consisting of three main stages, the A-Stage, B-Stage and C-Stage:

- **A-Stage**: The A-Stage is a low viscosity stage where resin and hardener have been combined but the rate of chemical reaction is noticeably low and the viscosity remains low.
- **B-Stage**: In the B-Stage, the rate of the chemical polymerisation reaction has increased dramatically and the resin becomes progressively more viscous. The reaction is exothermic and can be retarded by cooling to lower-than sub-zero temperatures.
- **C-Stage**: C-Stage resin is fully cured and heavily cross-linked.

Resins will have different cure cycles, with some resins completing cure at room temperature and others requiring heat. To optimise cross linking and consolidation, appropriate temperatures and dwell times need to be selected. A good cycle will optimise both the thermal and mechanical properties of a composite, reduce the presence of defects such as air bubble voids, minimise curing time, and will be cost effective. A basic cure cycle is shown in figure 3.9. In this figure, curves for curing temperature, viscosity, degree of cure and glass transition temperature (Tg) are shown, alongside their overlaps with respect to curing time.

Prepreg materials are a combination of fibres and B-Stage resin. Since the resin is in its B-Stage, prepreg is stored under sub-zero temperatures to retard the resin cure. Prepreg materials are easy to handle because they are in the B-Stage and the viscosity of the resin has risen enough to be (in combination with the fibres) semi-rigid. Prepreg materials can be cured using either autoclave or out-of-autoclave processes.

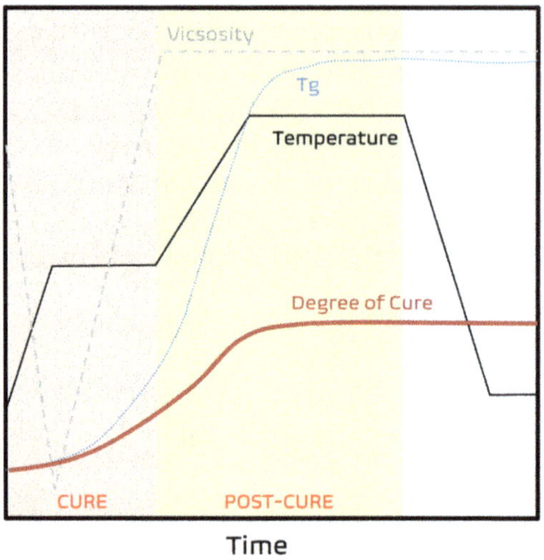

Figure 3.9. Simple representation of a cure cycle.

References

[1] Fei B 2018 High-performance fibers for textiles *Engineering of High-Performance Textiles* ed M Miao and H Xin (Woodhead: Duxford)

[2] Holmes M 2014 Global carbon fibre market remains on upward trend Reinf *Plast* **58** 38–45

[3] Giurgiutiu V 2015 *Structural Health Monitoring of Aerospace Composites* (Oxford: Academic)

[4] Cai Z and Gutowski T 1998 Consolidation techniques and cure control *Handbook of Composites* ed S T Peteres 2nd edn (Dordrecht: Springer)

[5] Cripps D, Searle T J and Summerscales J 2000 *Comprehensive Composite Materials Encyclopedia* ed R Talreja and J A Manson (Oxford: Elsevier) pp 737–61

[6] BS EN ISO 527-1:2019 *Plastics. Determination of Tensile Properties. General Principles* (British Standards Int.) p 34

[7] BS EN ISO 527-2:2012 *Plastics. Determination of Tensile Properties. Test Conditions for Moulding and Extrusion Plastics* (British Standards Int.) p 20

[8] BS EN ISO 527-3:2018 *Plastics. Determination of Tensile Properties. Test Conditions for Films and Sheets* (British Standards Int.) p 33

[9] BS EN ISO 527-4 *Plastics. Determination of Tensile Properties. Part 4. Test Conditions for Isotropic and Orthotropic Fibre-reinforced Plastic Composites* (British Standards Int.) p 33

[10] BS EN ISO 527-5:2009 *Plastics. Determination of Tensile Properties. Test Conditions for Unidirectional Fibre-reinforced Plastic Composites* (British Standards Int.) p 20

[11] ASTM D3039/D3039M-17 2017 *Standard Test Method for Tensile Properties of Polymer Matrix Composite Materials* (West Conshohocken, PA: ASTM Int.)

[12] ASTM D3410/D3410M-16e1 2016 *Standard Test Method for Compressive Properties of Polymer Matrix Composite Materials with Unsupported Gage section by Shear Loading* (West Conshohocken, PA: ASTM Int.)

[13] ASTM D695-15 2015 *Standard Test Method for Compressive Properties of Rigid Plastics* (West Conshohocken, PA: ASTM Int.)

[14] ASTM D6641/D6641M-16e1 2016 *Standard Test Method for Compressive Properties of Polymer Matrix Composite Materials Using a Combined Loading Compression (CLC) Test Fixture* (West Conshohocken, PA: ASTM Int.)
[15] ASTM D5528-13 2013 *Standard Test Method for Mode I Interlaminar Fracture Toughness of Unidirectional Fiber-Reinforced Polymer Matrix Composites* (West Conshohocken, PA: ASTM Int.)
[16] ASTM D7264/D7264M-21 2021 *Standard Test Method for Flexural Properties of Polymer Matrix Composite Materials* (West Conshohocken, PA: ASTM Int.)
[17] ASTM D2344/D2344M-16 2016 *Standard Test Method for Short-Beam Strength of Polymer Matrix Composite Materials and Their Laminates* (West Conshohocken, PA: ASTM Int.)
[18] Zhang W, Srivastava I, Zhu Y F, Picu C R and Koratkar N A 2009 *Small* **5** 1403–7
[19] Kunal S and Mrinal S 2013 Fiber crimp distribution in nonwoven structure *Front. Sci.* **3** 14–21

IOP Publishing

Composites Engineering: An A–Z Guide

Parvez Alam

Chapter 4

'D'

4.1 Damage models

Several damage models have been developed to predict failure in continuous fibre reinforced plastics (FRPs). Failure criteria are essentially mathematical predictions of stresses and strains at the onset of failure. Simple failure criteria consider one-dimensional maximum stresses, σ_{max}, and maximum strains, ε_{max}, and failure, f, is deemed to occur when these limits are reached, i.e. when either $f \geqslant \sigma_{max}$ or $f \geqslant \varepsilon_{max}$. The reality of composite failure is nevertheless significantly more complex and as such, several efforts have been made to improve the accuracy of composite damage models. The following lists a few damage models that are oft-used for modelling failure in FRPs.

4.1.1 Tsai–Hill

The Tsai–Hill failure criterion is based on Hill's failure criterion for anisotropic materials where ply rupture occurs according to equation (4.1), where σ_{11} and σ_{22} are normal stresses in 0° and 90° axes, respectively, γ_{12} is the shear stress on the 12 plane, X_{11} and X_{22} are the allowable strengths of the ply in 0° and 90° axes, and x_{12} is the allowable in-plane shear strength on the 12 plane. According to this failure criterion, failure occurs when index, $I_{TH} \geqslant 1$, and stresses in different directions affect failure simultaneously.

$$I_{TH} = \left(\frac{\sigma_{11}}{X_{11}}\right)^2 - \left(\frac{\sigma_{11}\sigma_{22}}{X_{11}^2}\right) + \left(\frac{\sigma_{22}}{X_{22}}\right)^2 + \left(\frac{\sigma_{12}}{x_{12}}\right)^2 \geqslant 1 \tag{4.1}$$

4.1.2 Tsai–Wu

The Tsai–Wu failure criterion predicts the onset of failure when failure index, $I_{TW} = 1$, equation (4.2), where F_i and F_{ij} are experimentally derived axial strength and biaxial strength parameters, respectively, and σ_i and σ_j are stresses where $(i, j = 1, 2, \ldots, 6)$.

$$I_{\text{TW}} = F_{ij}\sigma_i\sigma_j + F_i\sigma_i = 1 \tag{4.2}$$

In orthotropic materials, there is no longer coupling between normal and shear terms and the general form of Tsai–Wu is represented according to equation (4.3), where $F_1 = \sigma_{1t}^{-1} - \sigma_{1c}^{-1}$, $F_2 = \sigma_{2t}^{-1} - \sigma_{2c}^{-1}$, $F_3 = \sigma_{3t}^{-1} - \sigma_{13c}^{-1}$, $F_4 = F_5 = F_6 = 0$, $F_{11} = (\sigma_{1c}\sigma_{1t})^{-1}$, $F_{22} = (\sigma_{2c}\sigma_{2t})^{-1}$, $F_{33} = (\sigma_{3c}\sigma_{3t})^{-1}$, $F_{44} = (\tau_{23}^2)^{-1}$, $F_{55} = (\tau_{31}^2)^{-1}$, $F_{66} = (\tau_{12}^2)^{-1}$, and the subscripts t and c refer to tensile and compressive, respectively. The coefficients F_{12}, F_{13} and F_{23} can be obtained through equibiaxial testing, though these types of tests are proven to be extremely difficult.

$$\begin{aligned} &F_1\sigma_1 + F_2\sigma_2 + F_3\sigma_3 + F_4\sigma_4 + F_5\sigma_5 + F_6\sigma_6 + F_{11}\sigma_1^2 \\ &+ F_{22}\sigma_2^2 + F_{33}\sigma_3^2 + F_{44}\sigma_4^2 + F_{55}\sigma_5^2 + F_{66}\sigma_6^2 \\ &+ 2F_{12}\sigma_1\sigma_2 + 2F_{13}\sigma_1\sigma_3 + 2F_{23}\sigma_2\sigma_3 = 1 \end{aligned} \tag{4.3}$$

A further simplification can be made for plane stress analyses, equation (4.4).

$$F_{11}\sigma_1^2 + 2F_{12}\sigma_1\sigma_2 + F_{22}\sigma_2^2 + F_{66}\tau_{12}^2 + F_1\sigma_1 + F_2\sigma_2 = 1 \tag{4.4}$$

Failure modes are identified using the indices H_1, H_2 and H_3 where $H_1 = F_1\sigma_1 + F_{11}\sigma_1^2$, $H_2 = F_2\sigma_2 + F_{22}\sigma_2^2$, and $H_3 = F_66\tau_{12}^2$.

4.1.3 Hashin

The Hashin failure criterion aims to, semi-empirically, characterise tensile and compressive modes of failure of different phases of the composite (fundamentally, fibre and matrix). This failure criterion assumes that the failure modes of each phase occur on different fracture planes. The fibre failure mode in tension is shown in equation (4.5). The fibre failure mode in compression is shown in equation (4.6). The matrix failure mode in tension is shown in equation (4.7), and the matrix failure mode in compression is shown in equation (4.8). In these equations, S refers to strength and the subscripts T and C indicate tension and compression, respectively.

$$\left(\frac{\sigma_1}{S_{1T}}\right)^2 + \left(\frac{\sigma_{12}}{S_{12}}\right)^2 = 1; \quad \sigma_1 > 0 \tag{4.5}$$

$$\sigma_1 = -S_{1C}; \quad \sigma_1 < 0 \tag{4.6}$$

$$\left(\frac{\sigma_2}{S_{2T}}\right)^2 + \left(\frac{\sigma_{12}}{S_{12}}\right)^2 = 1; \quad \sigma_2 > 0 \tag{4.7}$$

$$\left(\frac{\sigma_2}{S_{21}}\right)^2 + \left[\left(\frac{S_{2C}}{2S_{21}}\right)^2 - 1\right]\frac{\sigma_2}{S_{2C}}\left(\frac{\sigma_{12}}{S_{12}}\right)^2 = 1; \quad \sigma_2 < 0 \tag{4.8}$$

4.2 Defects

Defects are often formed during the manufacturing process for fibre reinforced plastics. Common defects include but are not limited to; defective bonding, delamination, fibre misalignment, foreign particulates, micro- or macro-cracking in plies, and air-bubble voids. Each of these will be covered in more detail.

4.2.1 Blistering

Blistering is the raising and expansion of an area of a composite. Blistering is the result of changes in internal pressure close to the surface of the composite. Changes in pressure can arise through the emission of volatile gases, or from trapped moisture vapour.

4.2.2 Bonding defects and delamination

Bonding defects affect the entire range of composite materials, i.e.: particle reinforced composites (regular and irregular), fibre reinforced composites (both continuous and discontinuous), and structural composites (laminates, sandwich panels, and reinforced structures). The most common macro-scale defects observed in adhesively bonded composites and composite joints are delamination and debonding. Debonding occurs when the adhesive strength of an adhesive bond is inferior to the cohesive strength of the adherends. Debonding is instigated and/or catalysed by the presence of entrapped air, high concentrations of catalyst, low levels of adhesive, and cracks in either reinforcement or matrix.

4.2.3 Fibre defects

The primary types of fibre defect in composites are fibre breakage and fibre wrinkles. Fibres may be partially or fully broken prior to use and if a sufficient number of fibre defects exist, this can have significant effects on both the mechanical properties and the service life of a composite part. Fibre wrinkles arise during the manufacturing stage of an FRP and though fibre wrinkles obviously lessen the mechanical benefits of the wrinkled fibre, a large problem can be the effect the wrinkle has on surrounding matrix and fibres. Wrinkles can cause local fibre misalignment, give rise to resin rich regions, and may more easily accommodate foreign bodies, much to the detriment of the global mechanical properties of the composite.

4.2.4 Fibre misalignment

Fibre misalignment often occurs at the composite manufacturing stage. Two common reasons for fibre misalignment include low fibre–matrix dispersion and weak bonding between the fibres and the matrix. Fibre dispersion is important and is related to the fibre surface properties relative to the matrix properties. Ideally, fibre surfaces should have a good affinity to the matrix, however if fibres are hydrophobic and the matrix materials is hydrophilic, specific surface treatments may be needed to ensure fibre–matrix adhesion. Natural fibres are typically hydrophilic and may be subjected to hydrophobic surface treatments in order to increase the compatibility of the surfaces. Hydrophobic–hydrophobic interfaces are generally of good dimensional stability. Poor consolidation and matrices that are not fully cured are also causes of fibre misalignment. Misalignment of prepreg as well as mis-orientation of individual fibres can additionally result in fibre waviness defects, which affect the mechanical properties of fibre composites [1].

4.2.5 Foreign bodies

Foreign bodies can enter composites during the (pre-cured) manufacturing stage. These originate from a wide variety of sources including, broken fibre ends, grease from machinery, eroded backing film, and dust/particulates from the ambient surroundings.

4.2.6 Voids

A void is air that has been entrapped within a composite. Voids are detrimental to mechanical integrity and increases the statistical likelihood of early composite failure. Voids are often found between reinforcement plies and within bond-lines. They may also be confined to localised non-interconnected areas within the composite.

4.3 Draping and hot drape forming

Draping is essentially a technique for laying composite fabrics over a mould. The technique used then affects the intimacy of fabric contact with the mould, and changes in fibre orientation and direction are a result of conformation to the mould. Composite fabrics draped over a mould are important as they affect the ease of manufacture and ultimately therefore, the mechanical performance of composites moulded with curvatures. Fibre misalignment, buckling and increased waviness are outcomes of draping and of conforming to a corner angle or a curvature. Generally, the tighter the curvature the higher the expectation that there will be fibre misalignment and buckling.

The following methods are used to optimise the quality of prepreg draped by a ply-by-ply hand-lay process [2]:

- **Two handed guiding**: where prepreg is placed onto the (either male or female) mould using two hands.
- **One handed guiding**: where one hand is used to align prepreg material to a datum, while the second hand is used to lay the prepreg over the mould.
- **Manual folding**: where excess material is made easier to deal with by hand-folding the excess.
- **Tension-tension shearing**: where the prepreg material is sheared over the mould using two hands working in opposing directions.
- **Tension secured shearing**: where the prepreg material is held to an area of the mould by hand, while the other hand exerts a tensile pull on the prepreg to cause shearing and lay the prepreg tightly over the mould.

- **Smoothing with tension**: similar to tension secured shearing except the hand securing the prepreg is also used to smooth the prepreg over the mould surface (often using a smoothing tool).
- **Tool interaction shear forming**: where prepreg is pushed into a recess of the tool, which creates tension and allows the prepreg to shear.

As a general rule the tighter the corner angle, the greater the amount of time spent draping prepreg using the hand-lay process.

Hot drape forming involves heating up a flat prepreg laminated blank placed over a mould, after which a forming (vacuum) pressure is used to conform the material to the mould surface. The intimacy of composite conformity to the mould surface will depend on a number of factors including the thickness of the original blank, the heating temperature and pre-drape heating time applied to the blank, the tightness of mould corners, the level of complexity of the mould, and the forming pressure applied. While hot drape forming is a semi-automated process, it is still a faster route to part manufacture than the hand-lay process. Hot drape forming is commonly used in, e.g. the aerospace and automotive industries as out-of-autoclave processes.

4.4 Dry fibre material (prepreg)

Dry fibre materials are pre-impregnated with resin during manufacturing and are commonly referred to as prepreg materials, or prepregs. Common dry fibres are glass, carbon and Kevlar. Prepregs may be pre-impregnated with either thermoplastic or thermoset resins. Thermoset prepregs have superior mechanical properties to thermoplastic prepregs and these final properties affect their end-use. Thermoset prepregs additionally have a higher resistance to chemical corrosion and extreme temperatures, and as such while thermoplastic prepregs have been more typically used as automotive interiors, packaging containers and common home appliances, thermoset prepregs are used in many higher-end applications where mechanical strength is of greater importance.

4.4.1 Thermoset prepreg

Thermoset prepregs are cured to their B-stage and as such are tacky to touch. Though thermoset prepreg materials often come with a room temperature storage-life (often ca. 6 months), these materials can be stored at low temperatures (ca. $-20\ ^\circ$ C) to extend their storage-times and prevent C-stage curing (crosslinking) from taking place during storage. Activation of C-stage curing requires heat and pressure and as such, autoclaves are often used to cure prepregs, though they can of course also be cured using (vacuum) bagging for pressure, and ovens to supply heat.

4.4.2 General advantages of dry fibre material

There are several advantages to using dry fibre materials (prepregs) in place of wet-lay materials. Since prepregs are in their B-stage, the curing times of prepregs are lower than for wet-lay composites. Prepregs are also less wasteful than wet-lay composites, as (a) resin injected into the fibre network (rather than excess resin being wasted from wet-lay) and (b) both ply thickness and resin-to-fibre ratios are more easily controlled than in wet-lay processing. Prepregs are nevertheless more expensive and the use of specialist tools such as autoclaves require extra energy. They are nevertheless stiffer and stronger than wet-lay composites, are significantly lighter (thus raising the specific properties), and have far fewer post-cure defects.

4.4.3 Nanoengineered prepreg

In recent years, there have been efforts to incorporate nanomaterials into prepregs [3]. The four types of nanoengineered prepreg that are being researched (for the most part at low technology readiness levels) include: (1) two phase nanoengineered prepregs where nanomaterials are combined directly with the resin to form a nanomaterial prepreg, (2) three phase prepregs that are essentially dry fibre material but using a two phase nanoengineered prepreg as the resin that the dry fibres are pre-impregnated with (3) hybrid prepregs, which consist of both dry fibre and nano-material layers that are both pre-impregnated with resin and (4) nanomodified prepregs, which are dry fibre material (prepregs) combined with nanomaterial.

References

[1] Jusoh S M 2021 Manufacturing defects in natural fibre composites *Biocomposite Materials* ed M T H Sultan, M S A Majid, M R M Jamir, A I Azmi and N Saba (Berlin: Springer) pp 219–27

[2] Potter K and Ward C 2015 Draping processes for composites manufacture *Advances in Composites Manufacturing and Process Design* ed P Boisse (Woodhead: Cambridge) pp 93–109

[3] Islam A B M I and Kelkar A D 2017 Prospects and challenges of nanomaterial engineered prepregs for improving interlaminar properties of laminated composites—a review *MRS Commun.* 7 102–8

IOP Publishing

Composites Engineering: An A–Z Guide

Parvez Alam

Chapter 5

'E'

5.1 Ejection (demoulding)

Ejection, also called demoulding, is the process of mechanically removing cured composite from its mould. The process can be performed by hand, using tools, stripper plates, or using air pressure. A reason for why ejection is necessary is due to shrinkage around the mould. Shrinkage, S_i, is calculated according to equation (5.1), where D_{imp} is the dimension of the impression in direction i, and D_{part} is the part dimension in the same direction. As such, shrinkage is positive when the part experiences a negative strain. Shrinkage in the through thickness direction, S_z, can be calculated using equation (5.2), where B is the half thickness of the impression, z_s is the solid thickness, $C_m \cdot p_{\text{gf}}$ is the contribution to the relative thickness variation of the moulded part due to deformation (p_{gf} is the pressure distribution inside the impression at solidification, and C_m is a constant relating to mould rigidity). $\varepsilon_{z,\text{solid}}$ is the through thickness strain in the solid, and $\varepsilon_{z,\text{molten}}$ is the through thickness strain in the molten part, and these are calculated according to equations (5.3) and (5.4), respectively, where v is the specific volume and is a function of temperature and pressure, t_z is the onset time for thickness shrinkage, and $\dot{\varepsilon}_x$ and $\dot{\varepsilon}_y$ are the strains in orthogonal x and y axes expressed as a function of time, respectively [1].

$$S_i = \frac{D_{\text{imp}} - D_{\text{part}}}{D_{\text{imp}}} \tag{5.1}$$

$$S_z = -\frac{1}{B}\left(\int_0^{z_s} \varepsilon_{z,\text{solid}} \cdot dz + \int_{z_s}^{B} \varepsilon_{z,\text{molten}} \cdot dz\right) - C_m \cdot p_{\text{gf}} \tag{5.2}$$

$$\varepsilon_{z,\text{solid}} = \frac{1+v}{1-v}\int_{t_z}^{t}\left(\frac{1}{3 \cdot v}\right)dt - \frac{v}{1-v}(\dot{\varepsilon}_x + \dot{\varepsilon}_y)\Big|_{t_z}^{t} \tag{5.3}$$

doi:10.1088/978-0-7503-3527-0ch5

$$\varepsilon_{z,\text{molten}} = \int_{t_z}^{t} \left(\frac{1}{3 \cdot v} \right) dt \tag{5.4}$$

As a result of shrinkage, internal stresses develop in the part. Highly stressed parts are restricted in their movement during shrinkage by the mould and these require high ejection forces to remove. Resistance to ejection force, F_e, is primarily a function of the friction forces that develop through incomplete shrinkage and can be calculated using equation (5.5), where E is the Young's modulus of the cured composite, A is the total contact area between the mould and the cured composite, μ is the static coefficient of friction between the composite and the mould, α is the coefficient of linear expansion of the cured composite, Δ_t is the composite softening temperature less the mould tool temperature, d is the diameter of a circle with a circumference equal to the total projected perimeter of the mould, t is the wall thickness (average) and ν is the Poisson's ratio of the cured composite.

$$F_e = \frac{EA\mu\alpha\Delta_t}{\frac{d}{2t} - \frac{d}{4t}\nu} \tag{5.5}$$

If the angle, θ, of composite ejection under a given force is known, F_e can be simplified in its calculation, equation (5.6), where F_f is the friction force and can be determined using μ as $F_f = \mu \cdot F_n$ and F_n is the normal force perpendicular to angle θ.

$$F_e = F_f \cos\theta \tag{5.6}$$

5.2 Exotherm

An exotherm is a compound that releases heat during formation and intakes heat during decomposition. Resins mixed with hardeners or catalysts are examples of exotherms as they produce heat during gelation (gel point) where the compound increases in viscosity, which in turn exacerbates the rate of cross-linking during curing. Epoxy resins produce a lower exotherm (in general) than polyester resins. Heat produced in the exothermic reaction that takes place during gelation therefore affects the way in which a composite laminate cures as a function of thickness. There is hence importance in being able to predict heat transfer through a composite laminate, and to deduce the degree of cure.

Heat transfer through a composite laminate can be predicted in one-dimension according to equation (5.7), where C_c is the heat capacity of the composite (J (kg K)$^{-1}$), T is the temperature (K), t is time (s), r is the reaction rate (s^{-1}), w is the resin mass fraction, ΔH is the heat of reaction of resin (J kg^{-1}) and D_{therm} is the thermal diffusivity (m^2 s^{-1}).

$$\frac{\partial T}{\partial t} = D_{\text{therm}} \frac{\partial^2 T}{\partial x^2} + \frac{rw\Delta H}{C_c} \tag{5.7}$$

Thermal diffusivity in equation (5.7) describes how well the composite can spread heat and it takes into account both how quickly heat will flow through the composite (i.e. the thermal conductivity, k_c) and the speed at which the composite's temperature can change when heated (heat capacity, C). Since many composites are anisotropic, k_c is dependent on the direction in which it is measured or computed. In long fibre composites, the differences are significantly greater than in particulate composites and this is a result of the internal geometries and arrangements of the different composite phases. In long fibre uni-directional composites, thermal conductivity along the fibre axis, k_{axis}, is a fraction based calculation, equation (5.8), where v_m and v_f are the volume fractions of the matrix and fibre, respectively, and k_m and k_f are the thermal conductivities of the matrix and fibre, respectively.

$$k_{\text{axis}} = v_m k_m + v_f k_f \tag{5.8}$$

The through thickness thermal conductivity, k_z, can be calculated using a Springer–Tsai model, equation (5.9), where $B = 2\left(\frac{k_m}{k_f} - 1\right)$. Either k_{axis} or k_z can be used in place of k_c when a specific direction of a composite is under investigation.

$$\frac{k_z}{k_m} = \left(1 - 2\sqrt{\frac{v_f}{\pi}}\right) + \frac{1}{B}\left[\pi - \frac{4}{\sqrt{1 - \left(B^2 v_f/\pi\right)}}\right]\tan^{-1}\frac{\sqrt{1 - \left(B^2 v_f/\pi\right)}}{1 + B\sqrt{v_f/\pi}} \tag{5.9}$$

Thermal equilibrium in a composite is represented by equation (5.10), where H_u is the ultimate heat of reaction of the system and α is the degree of cure and $\alpha = \frac{\Delta H_t}{\Delta H_u}$, ($\Delta H_u$ is the total heat of reaction and ΔH_t is the heat of reaction with respect to time) [2].

$$\rho C_c \frac{\partial T}{\partial t} = \nabla k_c \nabla T + v_m \rho_m H_u \frac{\partial \alpha}{\partial t} \tag{5.10}$$

5.3 Extrusion

The extrusion process is a very old one that was originally used for metal forming in the late 1700s. It has developed considerably since then and is now used as a manufacturing method for a variety of composites including wood–plastic, metal–matrix and polymer–matrix composites. Fundamentally, the process involves pressing heated stock material through a die to form objects with fixed cross-sections. The die is thinner than the press and as such the material is partly compacted as it travels through the extruder. There is therefore, a size difference between the inlet and outlet cross sectional areas of an extruder and this difference is referred to as an extrusion ratio. The loading required to press material through from the inlet to the outlet increases as the extrusion ratio rises. The extrusion ratio should be chosen carefully as there are material limitations to extrusion. The extrusion ratios for hard and soft alloys for example, are 10:1–35:1 and 10:1–100:1, respectively.

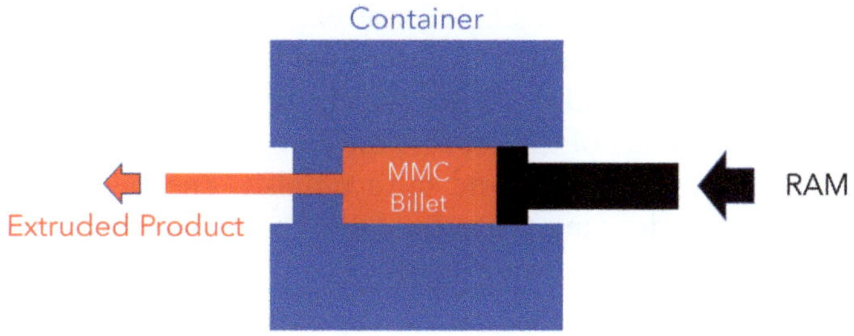

Figure 5.1. Example of direct extrusion of MMCs.

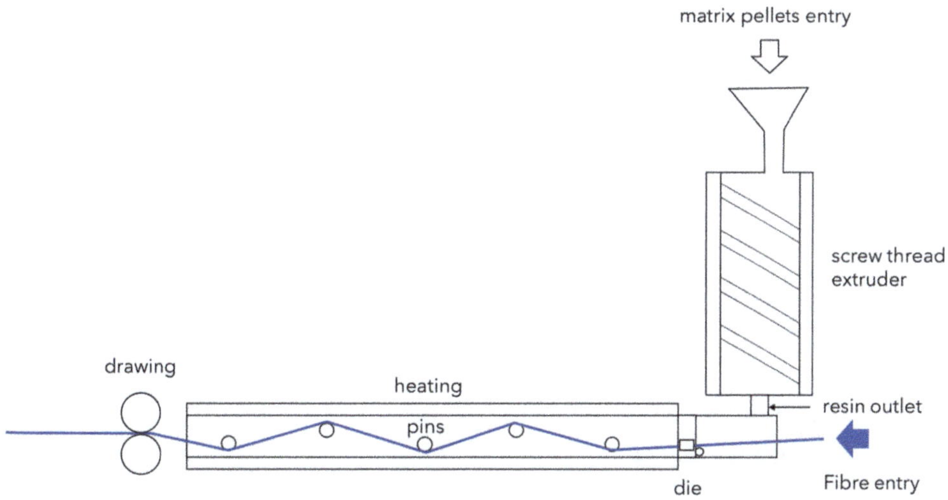

Figure 5.2. Example of extrusion of continuous fibre reinforced composites.

Metal matrix composites (MMCs) are commonly manufactured via the extrusion process. The process is noted to be beneficial to MMCs as discontinuous agglomerated reinforcements can break up into separate particles ensuring improved bonding to individual particulates or fibres. Porosity is also decreased during extrusion, which has mechanical benefits. Importantly, in discontinuous fibre MMCs (and FRPs), the extrusion process aligns fibres and creates directional mechanical properties. Reinforcement choice and the fraction used are important in the extrusion process as high aspect ratio reinforcements and/or high reinforcement fractions can cause clogging. Wood–plastic composites (WPCs) for example often focus on using wood flour or wood fibres/particles. Figure 5.1 shows a basic example of direct extrusion of MMCs, where a billet preloaded with reinforcing particulates is pressed through a die using a solid ram. A similar process for MMCs is indirect extrusion, where the ram is essentially an unmovable block, while the thin nozzle of the die itself presses through the reinforced billet. Continuous fibre reinforced composites are a little more complex as fibres need to be aligned and correctly oriented. Figure 5.2 shows a basic example of this process where matrix materials (e.g. polymer pellets) are fed into a single screw extruder and this is outlet as resin at a junction where continuous reinforcing fibre is fed into the system. The two mingle and are pressed through a small die, after which they are heated as they pass over rolling pins. At the final stage of the process the composite is drawn through rollers, ensuring compaction, consolidation and surface finish. The degree of consolidation will depend on the roller drawing pressure, while the surface finish will depend not only on this, but also on the surface properties of the rollers.

References

[1] Pontes A J V 2002 Shrinkage and ejection forces in injection moulded products *PhD Thesis* Universidade do Minho, Portugal

[2] Anandan S, Dhaliwal G S, Huo Z, Chandrashekhara K, Apetre N and Iyyer N 2018 Curing of thick thermoset composite laminates: multiphysics modelling and experiments *Appl. Compos. Mater.* **25** 1155–68

IOP Publishing

Composites Engineering: An A–Z Guide

Parvez Alam

Chapter 6

'F'

6.1 Fatigue

Fatigue is the primary mechanism of composite failure for structures subjected to cyclical loads. Cyclic loading results in the progressive damage of a composite, even when loading is within the limit of the composite elastic modulus. Under low-load levels (also: high-cycle fatigue), micro-damage accumulates and this can link-up into macro-damage, leading to catastrophic failure. The fatigue life, N, of a composite is the total number of loading cycles required to cause it to fail. There are different methods and approaches to fatigue testing in composites, the most common of which is to cyclically load the composite between two predefined levels of stress (a constant amplitude stress). The constant amplitude can be set as required and a loading that entails a complete reversal of cycling is termed tension–compression loading. Purely tensile cyclic loading at a constant amplitude (i.e. such that there is no negative loading) is termed tension–tension loading. Variable amplitude loading with irregularities of the cycling of stress is the least common method of testing as it is the least predictable and complex to analyse. Nevertheless, it can also be advantageous when real-world applications cannot be easily predicted using constant amplitude loading methods. An example is wind turbine blade technology, where blade loading is variable throughout its service life, and attempts are consequently being made to collect real-time loading data to improve the accuracy of the testing and modelling done on these structures.

6.1.1 Fundamental parameters of importance

A few important parameters involved in fatigue testing are provided below. The first is the stress range $\Delta\sigma = \sigma_{\max} - \sigma_{\min}$, which is essentially the difference between the maximum and minimum stresses imposed on a material. Another is the mean stress (σ_m), which is the average of the maximum and minimum values of stress and $\sigma_m = \frac{\sigma_{\max} + \sigma_{\min}}{2}$. The stress amplitude ($\sigma_a$) is also an important parameter as it indicates half the stress range and is represented as $\sigma_a = \frac{\Delta\sigma}{2} = \frac{\sigma_{max} - \sigma_{min}}{2}$. σ_a and $\Delta\sigma$ are always positive. This is because $\sigma_{max} > \sigma_{min}$, where tension is considered positive. Note though, that σ_{max}, σ_{min} and σ_m may be either positive or negative. The stress ratio R and the amplitude ratio A are commonly used parameters and are expressed as: $R = \frac{\sigma_{min}}{\sigma_{max}}$, and $A = \frac{\sigma_a}{\sigma_m}$, respectively.

6.1.2 ε–N curves

Generally speaking, as cyclic stress is increased, the fatigue life decreases. Plots for fatigue life are usually represented by either *S–N* curves or *ε–N* curves, where *S* represents stress, *ε* represents strain and *N* represents the number of cycles to failure, or, fatigue life. Figure 6.1 shows a generic *ε–N* curve. There are two distinct regions in this figure, one of which is represented by low-cycle fatigue (i.e. where the imposed cyclic stresses are high), and the other by high-cycle fatigue (i.e. where the imposed cyclic stresses are low). The low-cycle fatigue region is typically described by the Coffin–Manson law (linear damage law related to plasticity), which is mathematically represented as $\frac{\Delta\varepsilon_p}{2} = \varepsilon_f'(2N)^c$ where $\frac{\Delta\varepsilon_p}{2}$ is the plastic strain amplitude, ε_f' is an empirical constant (the fatigue ductility coefficient—i.e. the failure strain for a single reversal), $2N$ is the number of reversals to failure (N cycles) and c is an empirical constant (the fatigue ductility exponent). The high-cycle fatigue region is typically described by Basquin's law (a strain dependent law), $\frac{\Delta\varepsilon_e}{2} = \frac{\sigma_f'}{E}(2N)^b$, where $\frac{\Delta\varepsilon_e}{2}$ is the elastic strain amplitude, $\frac{\sigma_f'}{E}$ is the fatigue strength coefficient and b is a fatigue strength exponent. It should be noted that the fatigue life will increase as σ_f' increases

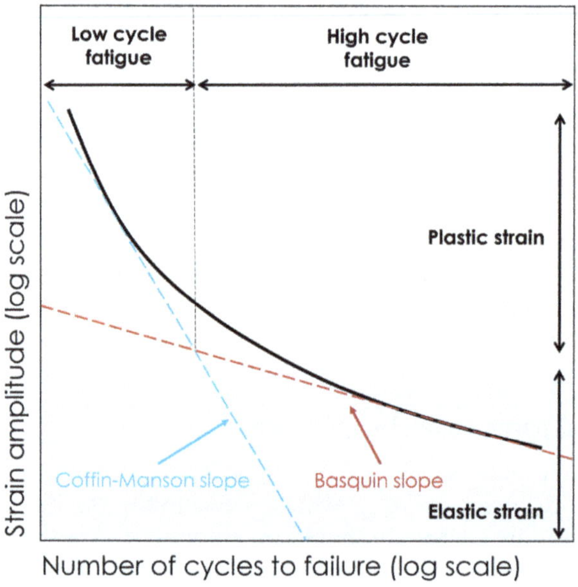

Figure 6.1. Generic example of an *ε–N* fatigue life curve.

and as b decreases. The fatigue life point ($2N_t$) transitions as $2N_t = \left(\frac{\varepsilon'_f E^{1/b-c}}{\sigma'_f}\right)$ and the total strain amplitude is the sum of both elastic and plastic strain: $\frac{\Delta\varepsilon}{2} = \frac{\Delta\varepsilon_p}{2} + \frac{\Delta\varepsilon_e}{2}$.

6.1.3 Damage evolution

Damage evolution in composites materials is from a mechanistic viewpoint, extremely complicated. This is because composites are inhomogeneous in their internal structures, are comprised of two or more phases, and have the additional complication of interfaces. The Paris law can be applied to some extent to describe the different stages of damage, and this law is based on the understanding that crack growth can be directly correlated to the stress intensity factor of the composite. The Paris law is generally, $\frac{dA}{dN} = A(\Delta K_I)^m$, where a is the crack length, N is the number of fatigue cycles, A and m are the Paris law coefficient and exponent, respectively, and ΔK_I is the oscillation of the stress intensity factor over a single load cycle.

Damage evolution can also be described pictorially, figure 6.2, where three stages are shown: I—intralaminar cracking including cracking in the matrix parallel to the fibre direction and/or cracking along fibre–matrix interfaces, II—delamination as a result of interlaminar shear stresses, and III—fibre or fibre ply breakage. The damage evolution of unidirectional continuous fibre laminates is shown in figure 6.3 at different stages over the lifetime of the composite [1].

6.2 Filler

Particulate fillers are often added to materials to reduce cost, and/or to improve properties. Paper for example, is a well-known composite made from wood pulp and

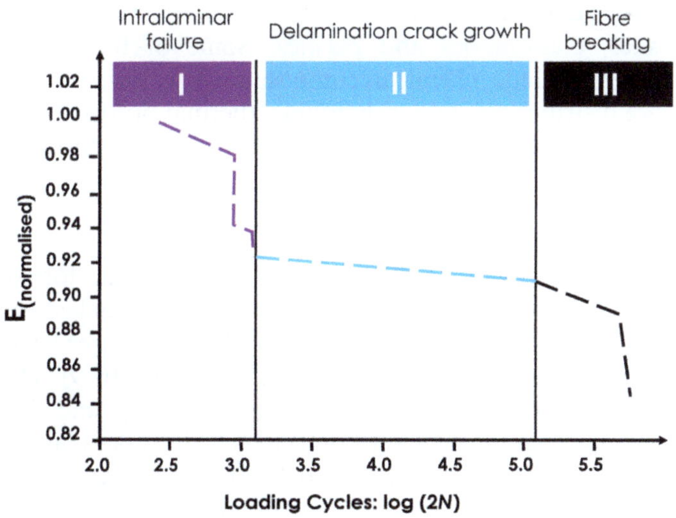

Figure 6.2. Fatigue damage evolution in composites.

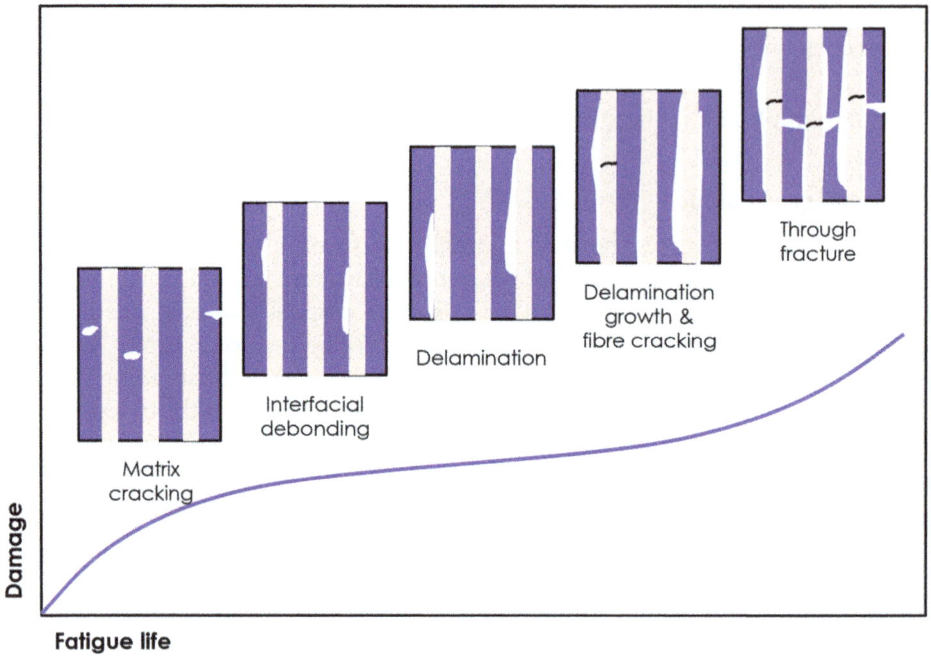

Figure 6.3. Damage modes and the characterisation of damaged states at different stages over the fatigue lifetime of a unidirectional composite.

starch. Up to 40% filler material such as talc or ground calcium carbonate can be added to paper to lower its cost since wood pulps tend to be considerably higher in cost. Additional benefits of adding mineral fillers such as these to base paper include improvements in opacity, printing, texture and surface finish after calendering. Fillers are not only commonly used in the paper industry, but are incorporated into dental composites, elastomerics and plastics. Some of the more sought after properties of fillers include chemical composition, density, hardness, refractive index, shape, size distribution, colour, hardness, thermal properties and wettability. Table 6.1 provides a details on general classes of fillers in composite materials [2].

6.3 Flitch

Flitch reinforcements are used for the most part in composites for construction. A flitch is a vertical reinforcing inserted into a structure, or attached to its sides. It is used to increase the rigidity of construction joints [3], to reinforce beams and trusses [4, 5], and to repair fractures or creep in structural elements [6–9]. The flitch can be either mechanically connected [9] or adhesively bonded [7], and though steel has been a traditional material for flitching, FRPs have gained popularity as flitches for both timber and concrete structures. Figure 6.4 shows examples of different types of internal adhesively bonded flitch joints, and figure 6.5 shows an example of internal placement of a flitch during a building repair scenario.

Table 6.1. Details on general classes of fillers in composite materials.

Filler	Description	Density (kg m^{-3})	Hardness (Moh's Scale)
Carbon black	Fine black carbon derived from the incomplete combustion of natural gas. Fine microaggregates with a 10–30 μm size distribution. Low refractoriness.	160–240	10
Kaolin (clay)	Kaolinites. White or cream coloured flat individual hexagonal platy particles as well as porous aggregates with a 0.5–50 μm size distribution. The particulates are soft and fine and have a high refractoriness.	320–640	<5
Bentonite (clay)	Mainly montmorillonite. Light cream coloured flat individual rounded hexagonal platy particles as well as porous aggregates. The particulates are soft and fine and have a high refractoriness.	800–960	5
Muscovite (mica)	Thin plates of white mica with high refractoriness.	190–320	2–5
Ground calcium carbonate	Sometimes integrates small quantities of magnesium carbonate. A blocky white particulate often described as rhombohedral. This mineral decomposes at high temperatures, is cheap, slightly alkaline and relatively soft. Soluble in acids.	640–960	3
Magnesium carbonate	Greyish blocky (rhombohedral) particulates that decompose at high temperatures and that are slightly soluble in acids.	640–960	3–5
Precipitated calcium carbonate	Chemically precipitated to form fine white particles with rounded structures forming into aggregates. These are finer than ground calcium carbonates and but decompose similarly at high temperatures. They are soluble in acids.	480–720	3
Talcs	These are either hydrous magnesium silicates or hydrous aluminium silicates. They are grey/green sometimes flaky particles with lamellar grains, and sometimes fibrous particles with fibres and splinters. They have low solubilities but high refractoriness.Hydrous magnesium silicates are denser and harder, while hydrous aluminium silicates are lower density and softer.	415–960	2–5
Crystalline silica	Crystalline silica such as quartz is white or colourless and is structured with irregular sharp angles to the particle. These particles are very hard and are generally inert. They are therefore insoluble and have a very high refractoriness.	880	5–7

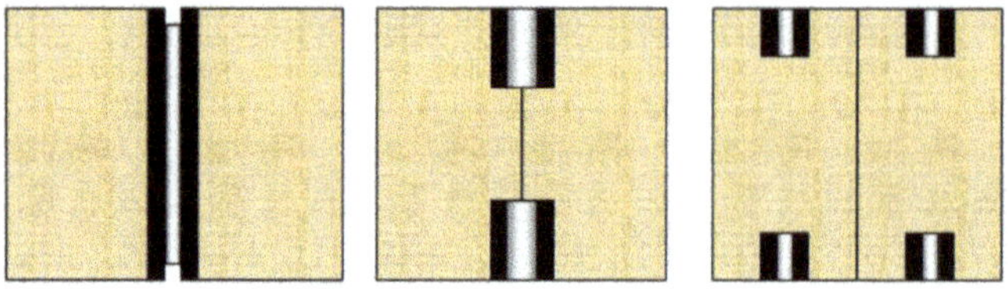

Figure 6.4. Examples of internal flitch reinforcing of timber beams showing (left) flitch running vertically through two parallel timber beams, (middle) flitch inserted into routed groover at the centre of two adhered timber beams and (right) adhered timber beams with small flitches inserted into routed grooves on the in the centre of each individual beam. Reprinted from [5] with permission from MDPI (CC BY 4.0).

Figure 6.5. Examples of internal composite reinforcing flitch repair of damaged timber beams from an historic timber structure on George Street, St. Albans, UK. Reprinted from [6], Copyright (2007), with permission from Elsevier.

A final nice example of an all-wood flitch joint is shown in figure 6.6, where both tension and compression joints were designed to enable the construction of the BOKU Vienna Robot Laboratory.

6.4 Fracture: interlaminar (FRP)

In any material, there are three fundamental modes of fracture usually referred to: tensile (mode I), shearing (mode II) and tearing/scissoring (mode III). In FRPs, these are normally considered in terms of interlaminar fracture. It should be noted that mode III is generally less commonly tested in FRPs than mode I and II fracture, as it is experimentally difficult to achieve a pure mode III delamination. As such, more detail has been afforded to understanding these more common modes of fracture in

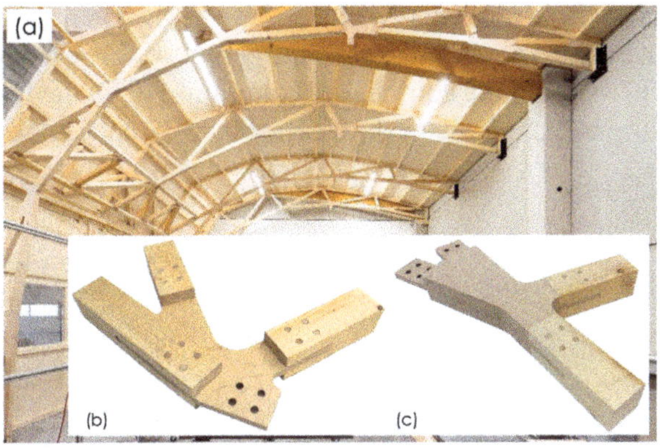

Figure 6.6. Examples of plywood flitch joints used in the construction of the BOKU Vienna Robot Laboratory: (a) the internal wooden structure, (b) tensile flitch joints used in the construction of (a), and (c) compression flitch joints used in the construction of (a). Reprinted from [10] with permission from MDPI (CC BY 4.0).

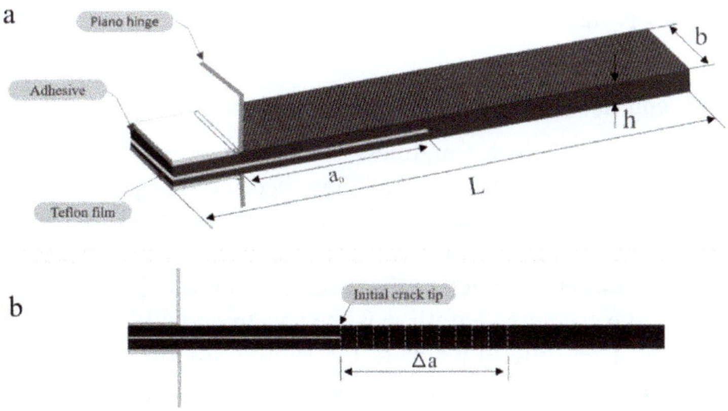

Figure 6.7. Schematic representation of a double cantilever beam (DCB) specimen: (a) 3D view (b) side view showing initial crack tip. Reprinted from [12] with permission from MDPI (CC BY 4.0).

composites. The reality of fracture in composites nevertheless, is considerably more complex than single mode fracture, and mixed modes of fracture should be expected. Fracture mechanics tests developed for composites are essentially focussed to an interlaminar plane and should be considered guidelines to build interpretation upon, rather than definitive transferable values to real-world applications.

6.4.1 Mode I: double cantilever beam (DCB) testing

Mode I interlaminar fracture toughness tests have been standardised in ASTM D5528-13, a standard that has been developed for unidirectional continuous fibre reinforced plastics [11]. It is a double cantilever set up and is shown in figure 6.7. As can be seen from the set up, it essentially involves a tensile loaded opening of two conjoined laminates. Fracture occurs at the interface between the laminates and crack opening occurs when the strain energy release rate G_I is equal to a critical value for strain energy release G_{IC}. This is calculated using one of three methods: the modified beam theory (MBT) method, the compliance calibration (CC) method, or the modified compliance calibration (MCC) method, equations (6.1)–(6.3), respectively.

$$G_I = G_{IC} = \frac{3P\delta}{2ba} \tag{6.1}$$

$$G_I = G_{IC} = \frac{nP\delta}{2ba} \tag{6.2}$$

$$G_I = G_{IC} = \frac{3P^2C^{\frac{2}{3}}}{2A_1bh} \tag{6.3}$$

In equations (6.1)–(6.3), P is the load, δ is the load point displacement, b is the specimen width, a is the delamination length, n is the slope of plot of log C versus log a, C is the compliance of the DCB specimen ($\frac{\delta}{P}$), A_1 is the slope of the plot of a/b against $C^{\frac{2}{3}}$, and h is the thickness of the DCB specimen.

6.4.2 Mode II: end notched flexure (ENF) testing

Mode II interlaminar fracture toughness tests have been standardised in ASTM D7905/D7905M-19e1, a standard that has been developed for unidirectional continuous fibre reinforced plastics [13]. It is an end notched 3-point bending method, as shown in figure 6.8. In this figure, a_0 is the initial distance of from an outer roller located under the crack to the crack tip, S is the distance between the two outer rollers, L is the specimen length, b is the specimen width and h is the specimen thickness. The test itself laterally shears the sample about its crack, which causes shear failure from and in line with the crack tip along the specimen span as a result of the stress concentration at the tip. Crack shearing occurs when the strain energy

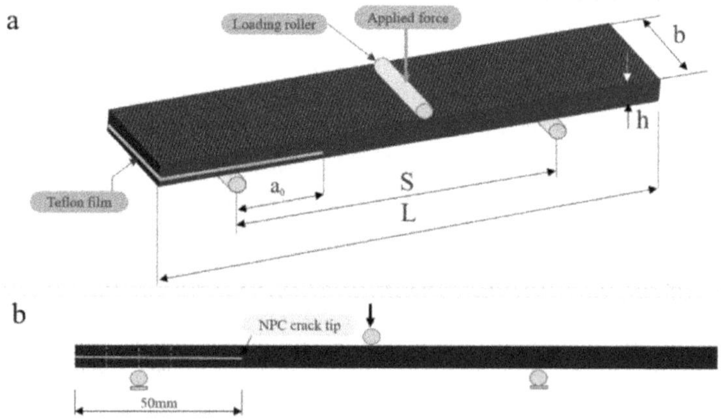

Figure 6.8. Schematic representation of an end notched flexure (ENF) specimen: (a) 3D view (b) side view showing initial crack tip. Reprinted from [12] with permission from MDPI (CC BY 4.0).

release rate G_{II} is equal to a critical value for strain energy release G_{IIC} and is shown in equation (6.4) for the configuration in figure 6.8.

$$G_{II} = G_{IIC} = \frac{9a^2P^2}{16E_{1,f}b^2h^3} \tag{6.4}$$

In equation (6.4), P is the applied force at fracture, a is the delamination length, and E is the flexural modulus of the specimen.

6.4.3 Mixed mode I and II: mixed-mode flexure (MMF) testing

Mixed mode I and II interlaminar fracture toughness tests have been standardised in ASTM D6671/D6671M-19, a standard that has been developed for unidirectional continuous fibre reinforced plastics [14]. Mixed-mode bending requires specialised testing equipment that not only applies a tensile pull to an existing crack (as with a DCB test) but simultaneously applies a three point bending load with in similitude with an ENF test. Figure 6.9 shows a schematic of the test set up, and additionally shows an example of a test fixture with a test specimen.

$$G_{IC} = \frac{12P_C^2(3c-L)^2}{16b^2h^3L^2E_{1,f}}(a_0+\chi h)^2 \tag{6.5}$$

$$G_{IIC} = \frac{9P_C^2(c-L)^2}{16b^2h^3L^2E_{1,f}}(a_0+0.42\chi h)^2 \tag{6.6}$$

$$G_{(I/II)C} = G_{IC} + G_{IIC} \tag{6.7}$$

Mixed mode I and II crack opening and shearing occurs when the strain energy release rate $G_{(I/II)}$ from both opening and shearing, equals a critical value for strain energy release $G_{(I/II)C}$, which is the sum of G_{IC} and G_{IIC}, equations (6.5)–(6.7). Here, χ is a crack correction parameter, P_C is the critical load of fracture, and c is the lever length.

6.4.4 Mode III: mixed-mode flexure (MMF) testing

The primary reason for why mode III fracture testing has received less attention than modes I and II fracture testing, is because it is extremely difficult to experimentally achieve a pure mode III delamination. It should also be noted that it is incorrect to assume that a mode III delamination is a shear delamination, since mode III tests that have been conducted on FRPs show that $G_{IIC} \neq G_{IIIC}$. With this in mind, a mode III delamination test should ensure that mode I and mode II is negligible or non-existent during the test, and that only small variations occur at most, in the mode III strain energy release rate G_{III}. While a number of tests have been proposed since the late 1980s, none of these have yet been shown to be able to completely eliminate the mode II behaviour within a mode III test. The most successful test to date seems to be the edge ring crack torsion (ERCT) mode III test, which was shown by Ge and co-workers [16] to enable pure mode III fracture while also exhibiting

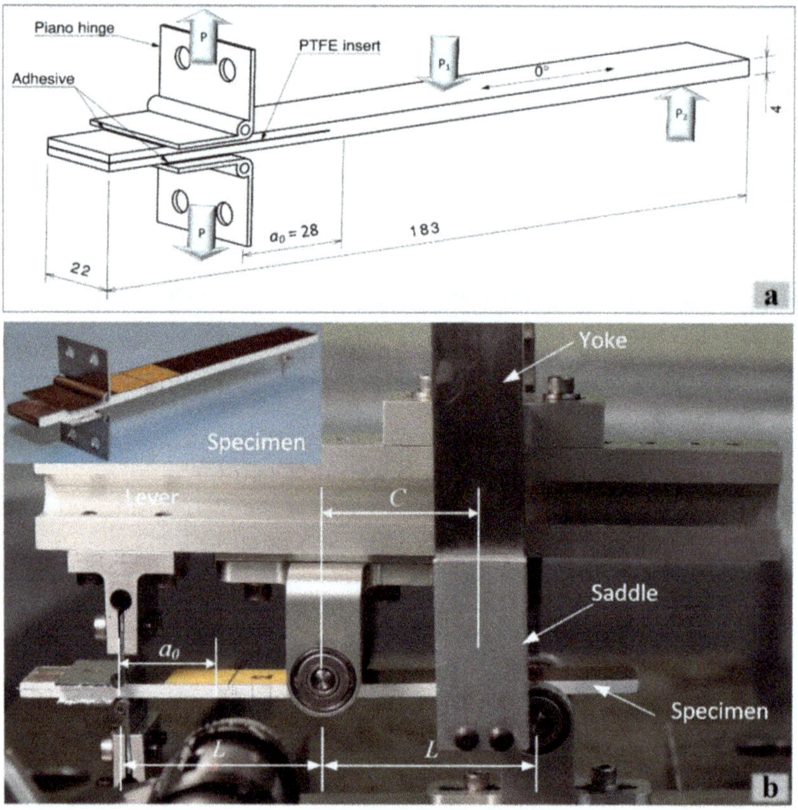

Figure 6.9. Mixed-mode bending (MMB) test specimen for mixed-mode I/II delamination: (a) scheme (b) example of a test fixture with a test specimen. Reprinted from [15] with permission from MDPI (CC BY 4.0).

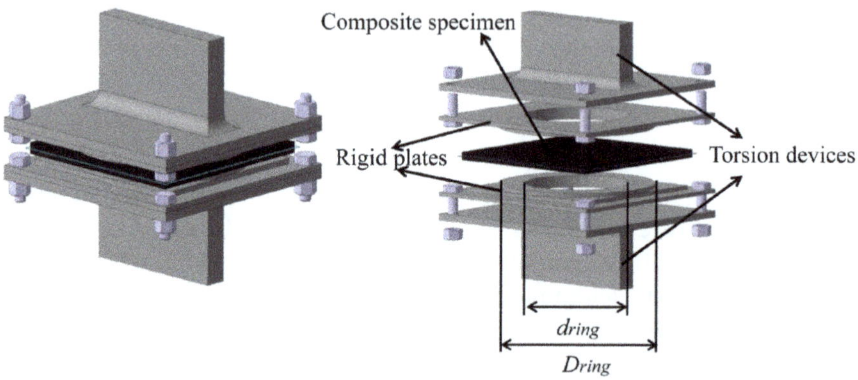

Figure 6.10. Example of an edge ring crack torsion test device (ERCT). Reprinted from [16], Copyright (2016), with permission from Elsevier.

only small variations of G_{III} along the crack front. The ERCT test is essentially a torsion test on a flat composite plate with rotation occurring about the z-axis of the plate. A crack ring is introduced around the edge of the specimen to localise mode III fracture to a specific plane. Figure 6.10 shows the basic construction of an ERCT test. While the ERCT test does have clear benefits in terms of mostly eliminating both mode I and II behaviours, a drawback is that the propagation of the crack is generally unstable.

References

[1] Alam P, Mamalis D, Robert C, Floreani C and Ó Bradaigh C M 2019 The fatigue of carbon fibre reinforced plastics—a review *Composites* **166** 555–79

[2] Cummins A B and Mulryan H 1937 *Industrial Minerals and Rocks: Nonmetallics Other Than Fuels* (American Institute of Mining and Metallurgical Engineers: Englewood, CO) Seeley W. Mudd Series

[3] Alam P and Ansell M P 2012 Short term lateral shear loading of nailed spruce-steel-spruce flitch joints and the modification of Eurocode 5 predictions for design resistance *J. Civil Environ. Eng.* **2** 1

[4] Alam P, Ansell M P and Smedley D 2013 Adhesively bonded plate and rod reinforcements for timber flexural beams *J. Int. Wood Prod.* **4** 52–61

[5] Alam P, Ansell M P and Smedley D 2012 Effects of reinforcement geometry on strength and stiffness in adhesively bonded steel-timber flexural beams *Buildings* **2** 231–44

[6] Smedley D, Alam P and Ansell M P 2007 George Street, St. Albans, UK; a case study in the repair of historic timber structures using bonded-in pultruded plates *Proc. of ACIC07 Advanced Composites in Construction, Bath, UK*

[7] Alam P 2014 The reinforcement of timber for structural applications and repair *PhD Thesis* University of Bath

[8] Alam P, Ansell M P and Smedley D 2009 Mechanical repair of timber beams fractured in flexure using bonded-in reinforcements *Composites* **B 40** 95–106

[9] Alam P and Ansell M P 2012 The effects of varying nailing density upon the flexural properties of flitch beams *J. Civil Eng. Res.* **2** 7–13

[10] Kromoser B, Braun M and Ortner M 2021 Construction of all-wood trusses with plywood nodes and wooden pegs: a strategy towards resource-efficient timber construction *Appl. Sci.* **11** 2568

[11] ASTM D5528-13 2013 *Standard Test Method for Mode I Interlaminar Fracture Toughness of Unidirectional Fiber-Reinforced Polymer Matrix Composites* (West Conshohocken, PA: ASTM Int.)

[12] Ali A, Andriyana A, Hassan S B A and Ang B C 2021 Fabrication and thermo-electro and mechanical properties evaluation of helical multiwall carbon nanotube-carbon fiber/epoxy composite laminates *Polymers* **13** 1437

[13] ASTM D7905/D7905M-19e1 2019 *Standard Test Method for Determination of the Mode II Interlaminar Fracture Toughness of Unidirectional Fiber-Reinforced Polymer Matrix Composites* (West Conshohocken, PA: ASTM Int.)

[14] ASTM D6671/D6671M-19 2019 *Standard Test Method for Mixed Mode I-Mode II Interlaminar Fracture Toughness of Unidirectional Fiber Reinforced Polymer Matrix Composites* (West Conshohocken, PA: ASTM Int.)

[15] Saadati Y, Chatelain J-F, Lebrun G, Beauchamp Y, Bocher P and Vanderesse N 2020 A study of the interlaminar fracture toughness of unidirectional flax/epoxy composites *J. Compos. Sci.* **4** 66

[16] Ge Y, Gong X, Hurez A and De Luycker E 2016 Test methods for measuring pure mode III delamination toughness of composite *Polym. Test.* **55** 261–8

Chapter 7

'G'

7.1 Gel coat

The purpose of a gel coat is essentially to protect the outer surface of a fibre reinforced plastic (FRP) composite, whilst concurrently improving its aesthetic finish. The gel coat also serves to decrease surface cracking by protecting it from environmentally induced microcracking, thereby increasing the longevity of the composite. Using a gel coat is also beneficial in reducing the need for post-manufacture processes such as painting or other forms of coating. The gel coat interface is important, as a well bonded interface will be long lasting and this in turn will influence the longevity of the composite. This section aims to introduce gel coat materials, the process and method of gel coating, and defects in gel coats alongside their methods of mitigation.

7.1.1 Gel coat material and lay-up

The gel coat material is a mixture of thermoset resin (e.g. epoxy or polyester) mixed with a hardener. The material is applied to the surface of the mould that an FRP composite will conform to, figure 7.1. Generally, since the hardener starts working swiftly, the gel coat may be applied as late as possible and as quickly as possible to the surface of the mould. Once the resin and hardener are mixed, the operator usually has about 15 minutes before the mixture starts to gel (harden), however the working time can be decreased or extended based on the concentration of catalyst added to the mixture. The gel coat is therefore applied in a viscous liquid state and crosslinks during curing. When this cross-linking occurs in conjunction with the cross-linking of the FRP matrix, an interphase can form between them, which may include surface bonding and polymer entanglements. The gel coat can fully cure within 2–4 hours, though longer cure-times are often preferred. Ideally, the final gel coat will be ca. 500 μm (500 g m^{-2}). A gel coat that is too thin will reveal the underlying fibre pattern, while an excessively thick coat may be prone to cracking.

doi:10.1088/978-0-7503-3527-0ch7

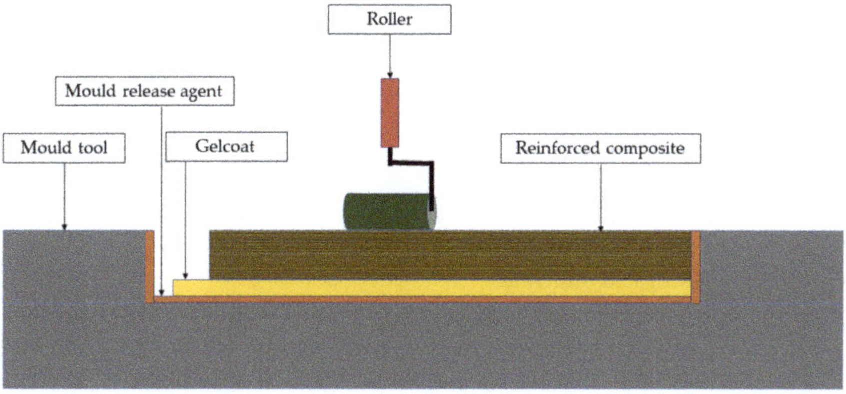

Figure 7.1. Location of gel coat in a composites manufacturing process. Reprinted from [1] with permission from MDPI (CC BY 4.0).

7.1.2 Comparison of manually applied and spray-coated gel coats

The gel coat is typically either manually applied (by brush or roller) or is sprayed onto the surface of the mould using a pressure pot or catalyst injection. Spraying the gel coat tends to provide more accurate and consistent coverage and the material is applied in layers, depending on the desired thickness of the final gel coat. Advantages of manually brushing the gel coat onto the mould include high air release and low styrene emission. The manual application methods are typically applied in 2–3 layers, each 2–300 microns in thickness. Manually brushing large areas of moulds is not entirely feasible, and in such cases rollers may be used. Spray coating is also a technique that can cover large mould areas. Spraying has to be optimised with respect to the viscosity of the coat and a primary concern with the method is the release of volatile organic compounds (VOCs) such as styrene during the spraying process. The emission of VOCs can be minimised by spraying at low pressures with high droplet sizes. The spray gun is positioned normal to the surface of the mould and spraying is carried out in two orthogonal directions at a distance of about 50 cm from the mould surface (though this is variable and is dependent on the nozzle size and the coverage area).

7.1.3 Gel coat defects and mitigation methods

A number of gel coat defects have been described in Polynt Composites' Application Guide [2]. These are summarised in table 7.1 below together with the main causes of the defects in gel coats, and measures of mitigation.

7.2 Glass fibres

Glass fibres are the oldest and most common fibre used in FRPs. They are used primarily as a substitute to heavier metallic components and materials, and though they are heavier than carbon fibres, they are able to strain more to failure, and are more tolerant to impact damage. This section aims to familiarise the reader with the manufacturing, types, properties and characteristics of glass fibres used in composites.

Table 7.1. A summary of possible gel coat defects, their causes and methods of mitigation.

Defect	Cause of defect	Mitigation/solution
Blistering (catalyst related)	Uncured gel coat, or catalyst remains unreacted	Improve catalyst mixing and addition concentration
Blistering (osmotic)	Arising due to air pockets, or water, solvent or oil instigated blistering	Optimise air release, rollout procedures and roller applicators
Cracking	Oversized (thickness) gel coat or impaction events	Checking of demoulding procedures and checking of possible opportunities for impact occurrences
Porosity	Resulting from poor air release and air entrapment	Optimise air pressure
Chalking	Poor condition of the mould, insufficient buffing	Solvent cleaning of buff area to deduce whether glossiness remains (no action required) or if glossiness has faded (buff needed)
Fish eyes	Impurities on mould such as dust or dirt	Clean and use lower viscosity coat
Alligatoring	Thin uncured gel coat	Check for overspray or leakage, check catalyst concentration

7.2.1 Manufacture of glass fibres

Glass fibres are manufactured in three basic stages: melting, forming, and drawing. In a direct melt process, the raw materials of silica and other selected metal oxides are heated to temperatures above 1200 °C in a furnace. If the silica is rolled into marbles while hot, after cooling, they can be packaged and transported, but must be remelted (the remelt process). Once either melted or remelted, the silica is drawn through a bushing (plate—often platinum—with many holes) at a specific drawing speed to obtain continuous fibre of a specific diameter after which it is wound upon a bobbin or collecting drum. The fibre size can be customised to suit its end-use and can be spun into different architectures accordingly.

7.2.2 Characteristics of glass fibres

Glass fibres are defined by their composition, which in turn affects their properties. There are several categories of glass fibre and the characteristics of each is described in more detail in table 7.2 [3].

7.2.3 Properties of glass fibres

Glass fibres are defined by their composition, which in turn affects their properties. There are several categories of glass fibre and the properties of each is provided in table 7.3 [3].

Table 7.2. Characteristics of different types of glass fibre.

Type	Description
A-glass	A-glass (alkali glass or soda lime glass) is the most common type of glass fibre. It is chemically stable, hard and easy to work with. A-glass has high electrical resistivity and strength, is a recyclable glass fibre (easily remelted), and is made up of silica, alumina, sodium carbonate, dolomite and lime.
AR-glass	AR-glass (alkali resistant glass) is alkali resistant due to the presence of zirconia. It is used in concrete, providing strength and flexibility to retard cracking in the concrete.
C-glass	C-glass (chemical glass) has high corrosion and chemical resistance as it contains high fractions of calcium borosilicate. It is often applied as a surface laminate on composites in aqueous and chemical environments such as pipes and containers.
D-glass	D-glass has a low dielectric constant as it contains boron tri-oxide and a low thermal expansion as it contains borosilicate. It is generally used in optical cables and electrical appliances.
E-glass	E-glass (electrical glass) is a high strength glass with high electrical resistivity. It is one of the main glass fibres used as reinforcement in composite materials. It is an alkali glass containing silica, alumina and other metal oxides (calcium, magnesium, boron, sodium and potassium). It is attractive as it is low cost, has excellent mechanical properties, high chemical resistance, and is electrically insulating. It is widely used in composites applied to the aerospace, automotive, marine and renewable energy technology industries.
ECR-glass	ECR-glass (electronic glass) contain calcium aluminosilicates that give it the properties of high strength, electrical conductivity, and corrosion resistance to acids. It has an excellent performance-cost ratio.
S-glass	S-glass (also R-glass or T-glass) are very high strength glass fibres used in the aerospace industries and in military ballistic technologies. These fibres are drawn with small fibre diameters and are relatively high cost.
S_2-glass	S_2-glass (silica-glass) is a high performance glass fibre with high fractions of silica. It excels in terms of its specific properties, its temperature resistance, its impact damage tolerance and its performance-cost ratio.

Table 7.3. Properties of different types of glass fibre, where ρ—density, σ_t is tensile strength, E is Young's modulus, ϵ is elongation, CTE is the coefficient of thermal expansion, ν is Poisson's ratio and RI is the refractive index.

Type	ρ **(g cc^{-1})**	σ_t (GPa)	E **(GPa)**	ϵ (%)	CTE (10^{-7}/°C)	ν	**RI**
A-glass	2.44	3.31	68.9	4.8	73		1.583
AR-glass	2.7	3.24	73.1	4.4	65		1.562
C-glass	2.52	3.31	68.9	4.8	63		1.533
D-glass	2.14	2.42	51.7	4.6	25		1.465
E-glass	2.58	3.44	72.3	4.8	54	0.2	1.558
ECR-glass	2.72	3.45	80.3	4.8	59		1.579
S-glass	2.54	4.14	85.5	4.8	33		1.546
S_2-glass	2.46	4.89	86.9	5.7	16	0.22	1.521

7.3 Glass transition

The glass transition temperature (Tg) is the temperature, or the range of temperatures, at which an amorphous material changes from its glassy state to its rubbery state. The Tg is common in polymers, and as such, polymer matrix composites (PMCs) will also exhibit a Tg, which is a function of not only the polymer matrix, but additionally the interface between the polymer matrix and the fibres. The service conditions of PMCs should generally remain well below the Tg of a load bearing composite if the material is to preserve its structural integrity. If PMCs are used in service conditions above the Tg of composite, the polymer will become rubbery and although cross-linking of polymer molecules in thermosetting plastics will prevent molecular slip, the polymer as a whole will be softened (rubbery) and will be unable to bear its service loads.

Since the Tg is related to the polymer component of the composite, and to polymer–reinforcement interfaces, the polymer and its interfaces play a vital role in any situations where the mechanical behaviour is both matrix-dominated and at an elevated temperature. Regardless of whether the Tg has been surpassed or not, the mechanical properties of composites subjected to matrix-dominated load bearing are reduced as temperature is increased, and this is most evident at the Tg. For example, PMCs subjected to temperatures at or above the Tg can be reduced in stiffness by as much as 3-fold.

There are three main material properties by which means Tg can be determined for thermosetting matrix PMCs. These include the heat capacity, the coefficient of thermal expansion (CTE) and the modulus. Each will be detailed individually.

7.3.1 Heat capacity method—differential scanning calorimetry (DSC)

This method focusses on increasing the heat capacity of a composite, and observing the change in heat required to cause the increase. A differential scanning calorimeter (DSC) is used and the method can be useful for composites when there is a large increase in the measured heat capacity of a PMC, as this can be associated with the Tg of the PMC. It should be noted that the method works better when there is more polymer available in the sample. As the fraction of reinforcement in a PMC increases, the decrease in the fraction of the polymer component makes detection of the heat capacity difference harder, and this difficulty increases as a function of increasing reinforcement fraction. The method should therefore preferably be used with PMCs containing a high enough fraction of polymer. The heat capacity method is standardised in ASTM 7426 [4]. As a composite cures, the change in heat capacity reduces such that a fully cured composite will show no variation in heat flow. The method can therefore also be used to detect curing. Figure 7.2 shows an example scheme of a DSC curve indicating regions associated with the Tg and with curing (due to prolonged exposure to high heat). The curing curve shows up as more shallow when the pre-DSC cure level in the sample is higher.

7.3.2 CTE method—thermomechanical analyser (TMA)

The CTE method involves measuring a physical change in the dimension of a sample with respect to a temperature increase applied to the sample. A high dimensional

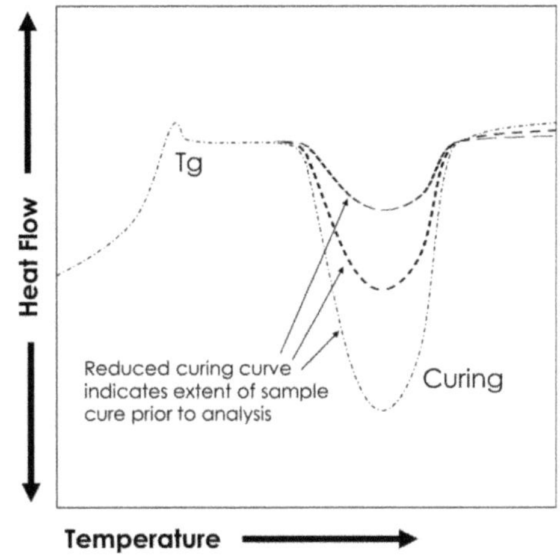

Figure 7.2. Example scheme of a DSC curve indicating regions associated with the Tg and with curing (due to prolonged exposure to high heat). The curing curve is more shallow if the cure level in the sample is higher.

increase is usually associated with the Tg in polymers and thus also PMCs. A thermomechanical analyser (TMA) is usually used to enable the measurement and this instrument fundamentally measures a change in the length of a specimen as a function of a temperature increase. The temperature is normally applied slowly so that the effect is not localised to PMC surfaces. The method is standardised in ASTM E1545 [5]. The expansion method described is one of two TMA methods, the second main alternative method being a penetration method (essentially a hardness measurement), also described in this standard. In the penetration method, the Tg is understood to correlate with a large change in penetration. A flexural TMA method may also be used, and in similitude with the penetration method, a large increase in flexural displacement is associated with the PMC Tg.

7.3.3 Modulus method—dynamic thermal mechanical analyser (DMTA)

The modulus of a PMC under mechanical loading is an oft-used method for the determination of Tg. Essentially, during mechanical loading, large changes in the modulus of the material can be associated with the Tg. A dynamic mechanical thermal analyser (DMTA) is a useful tool for this purpose, as it can apply an (oscillated) harmonic potential to the sample whilst recording the displacement at different temperatures that can be ramped at different rates. There are several standards that are applicable to PMCs, some focussing on tensile loading such as the ASTM D5026 [6], some on compressive loading such as the ASTM D5024 [7], some focussed on three-point flexure such as the ASTM D5023 and ASTM D7028 [8, 9], dual cantilever beam flexure (ASTM D5418 and ASTM D7028) [9, 10] and standards associated with torsion (ASTM D5279) [11]. The ASTM standard D4065 [12] provides information on methods by which means Tg can be calculated from a DMTA test. The method used should consider the type and style of reinforcement in the PMC. A PMC reinforced with particulates is close in behaviour

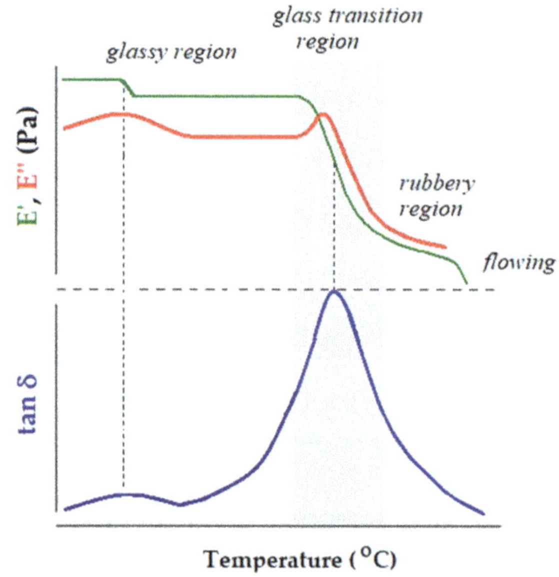

Figure 7.3. Example scheme of a DMTA curve indicating the region associated with the Tg, which is normally taken from the tan–δ peak. Reprinted from [13] with permission from MDPI (CC BY 4.0).

to an isotropic material and can be tested using any of the above standards. However, the stiffness in a unidirectional stiff fibre reinforced plastic may hinder the accuracy of most test methods and simply supported three-point flexure is advisable over any cantilever option as the fixed end of a cantilever will overshadow the dynamic mechanical effects of the polymer matrix and their interfaces. The Tg in a DMTA test is normally taken at the peak of the tan–δ curve. Figure 7.3 provides an informative scheme of DMTA curves including the storage modulus (E'), the loss modulus (E'') and the tan–δ peak, which is associated with the material Tg.

References

[1] Pomazi A and Toldy A 2019 Multifunctional gelcoats for fiber reinforced composites *Coatings* **9** 173

[2] Polynt Composites Application Guide 2015 Polynt Composites USA Inc., 99 East Cottage Avenue Carpentersville, IL 60110, USA [http://www.polynt.com]–Accessed 05.06.2021

[3] Sathishkumar T P, Satheeshkumar S and Naveen J 2014 Glass fibre-reinforced plastics–a review *J. Reinf. Plast. Compos.* **33** 1258–75

[4] ASTM D7426-08(2013) 2013 *Standard Test Method for Assignment of the DSC Procedure for Determining Tg of a Polymer or an Elastomeric Compound* (West Conshohocken, PA: ASTM Int.)

[5] ASTM E1545-11(2016) 2016 *Standard Test Method for Assignment of the Glass Transition Temperature by Thermomechanical Analysis* (West Conshohocken, PA: ASTM Int.)

[6] ASTM D5026-15 2015 2015 *Standard Test Method for Plastics: Dynamic Mechanical Properties: In Tension* (West Conshohocken, PA: ASTM Int.)

[7] ASTM D5024-15 2015 2015 *Standard Test Method for Plastics: Dynamic Mechanical Properties: In Compression* (West Conshohocken, PA: ASTM Int.)
[8] ASTM D5023-15 2015 *Standard Test Method for Plastics: Dynamic Mechanical Properties: In Flexure (Three-Point Bending)* (West Conshohocken, PA: ASTM Int.)
[9] ASTM D7028-07(2015) 2015 *Standard Test Method for Glass Transition Temperature (DMA Tg) of Polymer Matrix Composites by Dynamic Mechanical Analysis (DMA)* (West Conshohocken, PA: ASTM Int.)
[10] ASTM D5418-15 2015 *Standard Test Method for Plastics: Dynamic Mechanical Properties: In Flexure (Dual Cantilever Beam)* (West Conshohocken, PA: ASTM Int.)
[11] ASTM D5279-13 2013 *Standard Test Method for Plastics: Dynamic Mechanical Properties: In Torsion* (West Conshohocken, PA: ASTM Int.)
[12] ASTM D4065-20 2020 *Standard Practice for Plastics: Dynamic Mechanical Properties: Determination and Report of Procedures* (West Conshohocken, PA: ASTM)
[13] Cristea M, Ionita D and Iftime M M 2020 Dynamic mechanical analysis investigations of PLA-based renewable materials: how are they useful? *Materials* **13** 5032

IOP Publishing

Composites Engineering: An A–Z Guide

Parvez Alam

Chapter 8

'H'

8.1 Halpin–Tsai model

The Halpin–Tsai model is a method for predicting the elastic modulus of composites. It is applicable to particulate reinforced composites, and to the transverse axis of continuous fibre reinforced composites. It is different to other fundamental predictive modulus models (such as the rule of mixtures and the transverse rule of mixtures) as it is a semi-empirical model that takes into account the geometry of the reinforcement. The generic structure of the Halpin–Tsai model is shown in equation (8.1), where E_t is the composite transverse Young's modulus, E_m is the Young's modulus of the matrix material, V_r is the volume fraction of reinforcement, ξ is a shape specific parameter for the reinforcement, and η is calculated using equation (8.2), where E_r is the Young's modulus of the reinforcement.

$$E_t = \frac{E_m\,(1 + \xi\eta V_r)}{(1 - \eta V_r)} \tag{8.1}$$

$$\eta = \frac{\dfrac{E_r}{E_m} - 1}{\dfrac{E_r}{E_m} + \xi} \tag{8.2}$$

$$\xi = 2\left(\frac{a}{b}\right) \tag{8.3}$$

For fibres or particles with either approximately circular or square cross-sections (i.e. where there is an aspect ratio close to unity), $\xi = 2$. ξ can be determined according to equation (8.3), where a is the dimension of the reinforcement in the direction of loading, while b is the dimension of the reinforcement transverse to the

doi:10.1088/978-0-7503-3527-0ch8

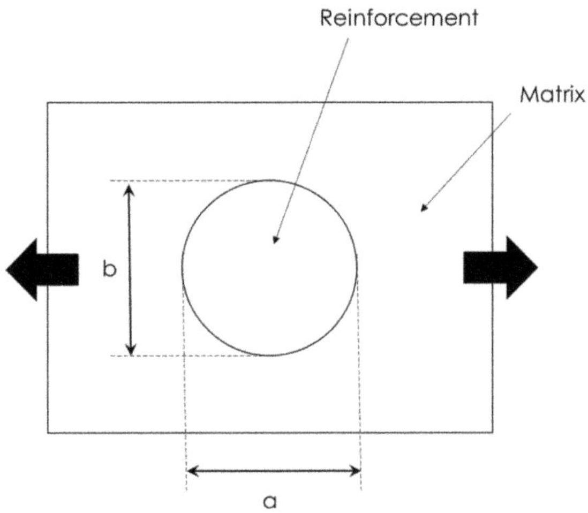

Figure 8.1. Method for the determination of ξ from a single reinforcement.

direction of loading, figure 8.1. Obtaining accurate values for ξ can be difficult since particle and fibre diameter size distributions can sometimes be very wide.

8.2 Hygroscopy

8.2.1 Isotropic hygroscopy

Hygroscopy is the ability for a material to absorb water. Certain materials, such as polymers, will swell as a result of hygroscopic water intake and this expansion can be understood by comparing the relative changes in dimension against the relative changes in mass as a function of water absorption into a material. This is termed the coefficient of hygroscopic expansion (CHE) or the coefficient of moisture expansion (CME), and is described by equation (8.4), where l is the length and m the mass of the water absorbing material. This equation is fine provided the material is isotropic and homogeneous. Neat polymers usually fall into this category and as such this equation can be used for neat polymers. In the case of composites, the equation is valid as an approximation for composites with global isotropic properties and in which the reinforcements (of aspect ratio close to unity) are homogeneously distributed through the matrix.

$$\mathrm{CME} = \frac{\frac{\Delta l}{l}}{\frac{\Delta m}{m}} \tag{8.4}$$

8.2.2 Micromechanical CME calculations for unidirectional laminates

Most composites are anisotropic and their properties are affected primarily by their reinforcement architectures. In a unidirectional continuous fibre reinforced plastic, CME can be approximated to three directions of orthogonal symmetry, which are associated with their longitudinal, CME_l, transverse, CME_t and thickness, CME_{th} directions. For the sake of simplification, here we consider that $CME_t \approx CME_{th}$ and the CME_l and CME_t can be calculated using equations (8.5) and (8.6), respectively, where E_f and E_m are the Young's moduli of fibre and matrix, respectively, V_f and V_m are the volume fractions of fibre and matrix, respectively, CME_f and CME_m are the coefficients of moisture expansion for fibre and matrix, respectively, E_l is the longitudinal Young's modulus of the unidirectional FRP, ν_f and ν_m are the Poisson's ratios for fibre and matrix, respectively, CME_l is the coefficient of moisture expansion in the longitudinal direction of the composite and ν_{lt} is the Poisson's ratio of the composite corresponding to a contraction in the transverse direction when under an extension in its longitudinal direction. These equations work if the fibres are isotropic and CME is equal in the fibres in all directions.

$$CME_l = \frac{\left(E_f V_f CME_f + E_m V_m CME_m\right)}{E_l} \tag{8.5}$$

$$CME_t = \left(1 + \nu_f\right) V_f CME_f + \left(1 + \nu_m\right) V_m CME_m - CME_l \nu_{lt} \tag{8.6}$$

8.2.3 Micromechanical CME calculations for laminates containing orthohygroscopic fibres

Certain fibres are orthohygroscopic in nature, exhibiting different CME in different directions of orthogonal symmetry. Natural fibres are example of fibres showing orthohygroscopic properties and this is due to the highly oriented nature of the cellulose molecules. High directionality synthetic fibres such as Kevlar are also known to have orthohygroscopic properties. The CME_t of a composite containing orthohygroscopic fibres in an isotropic matrix can be calculated using equation (8.7), where the subscript $_{f,l}$ refers to fibre in the longitudinal axis, $_{f,t}$ refers to the fibre in its transverse axis, and $_{f,lt}$ refers to the fibre contracted in its transverse direction while under and extension in its longitudinal direction, α is the longitudinal Young's modulus and is calculated $\alpha = E_{f,l} V_f + E_m V_m$ and β is the product of modulus and CME in the longitudinal direction and is calculated $\beta = E_{f,l} V_f CME_{f,l} + E_m V_m CME_m$.

$$\begin{aligned} CME_t = &\left(1 + \frac{CME_{f,l}}{CME_{f,t}}\right) V_f CME_{f,t} + (1 + \nu_m)\, CME_m \\ &- \left(\nu_{f,lt} V_f CME_f + \nu_m V_m\right)\right) \frac{\alpha}{\beta} \end{aligned} \tag{8.7}$$

8.2.4 Micromechanical CME calculations for laminates containing non-absorbing fibres

Several engineering fibres are impermeable and do not swell when in water. Examples include glass, boron and graphite fibres. In such cases, the CME_l and CME_t can be calculated (assuming the fibres are isotropic) using equations (8.8) and (8.9), respectively.

$$\mathrm{CME}_l = \frac{E_m V_m \mathrm{CME}_m}{E_l} \tag{8.8}$$

$$\mathrm{CME}_t = (1 + \nu_m)\, V_m \mathrm{CME}_m - \mathrm{CME}_l \nu_{\mathrm{lt}} \tag{8.9}$$

$$\mathrm{CME}_t = \left(1 + \nu_m\right) V_m \mathrm{CME}_m - \nu_m V_m \left(\frac{E_m V_m \mathrm{CME}_m}{E_{f,l} V_f + E_m V_m}\right) \tag{8.10}$$

If the fibres are orthotropic, CME_t can be calculated according to equation (8.10).

8.3 Hybrid composite

Hybrid composites are composites containing more than one type of reinforcement material. Common combinations are glass with carbon fibres and different types of natural fibres in combination. By using more than one type of composite reinforcement, designers are able to increase the extent to which they can control and manipulate the properties of composites to those that are desirable for specific applications. Cost can also play a major role in the choice to combine different reinforcement types. Using carbon and glass hybrids as an example in this section, carbon is a high-cost high-performance engineering fibre, while glass is a lower cost fibre that is weaker and less stiff than carbon. Combining the two results in a reasonably costed composite with superior properties to glass fibre reinforced plastic. The decision to combine glass and carbon fibre may additionally be to improve fracture behaviour. Carbon fibres are more brittle under impact than glass, and combining both glass and carbon leads to the creation of a composite with improved tolerance to impact damage, whilst also saving on cost. A drawback of replacing carbon fibre with glass fibre is of course weight, and as such, each of the above mentioned factors, alongside others (e.g. electrical conductivity), need to be balanced out during the design process.

A few examples of hybrid glass-carbon fabric architectures are shown in figure 8.2(a) from [1]. Here, unidirectional interplay hybrids are shown against reference carbon and glass sheets. The hybrids are one, two and three carbon-glass tow repeats and the images are magnified for one of the samples in figure 8.2(b).

A hybrid glass-carbon composite generally exhibits shared mechanical behaviour that reflects the properties of the different constituent reinforcement phases, figure 8.3(a). The variation in properties is nevertheless not necessarily linear with respect to the volume fraction of reinforcement, as shown in figure 8.3(b).

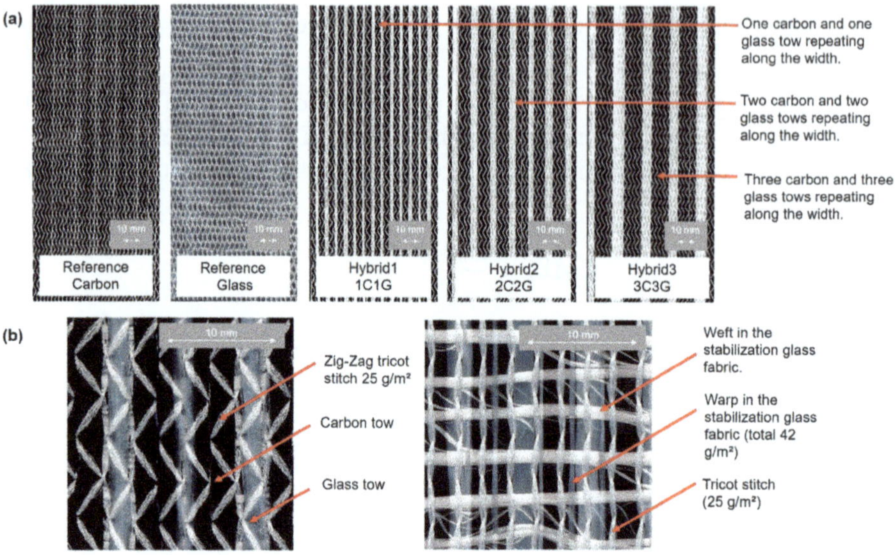

Figure 8.2. Photographs of developed fabric reinforcements: (a) reference and hybrid fabrics; and (b) magnified images of the face and back of a Hybrid1 fabric. Reprinted from [1] with permission from MDPI (CC BY 4.0).

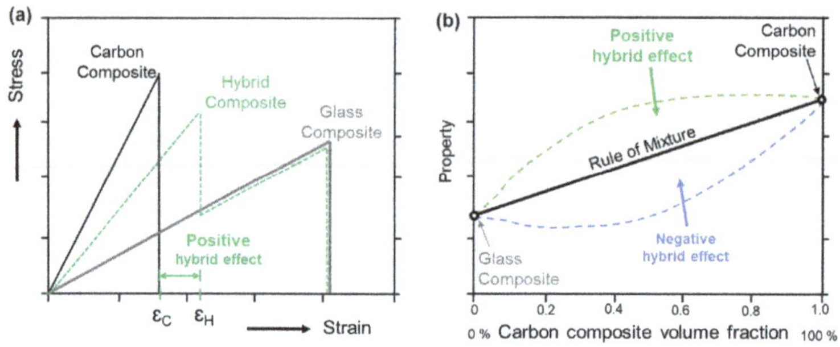

Figure 8.3. Illustration of the definitions of the hybrid effect or synergistic effect for C/G hybrid composite, based on: (a) apparent enhancement in failure strain; and (b) deviation from Rule of Mixture. Reprinted from [1] with permission from MDPI (CC BY 4.0).

Reference

[1] Rajpurohit A, Joannes S, Singery V, Sanial P and Laiarinandrasana L 2020 Hybrid effect in in-plane loading of carbon/glass fibre based inter- and intraply hybrid composites *J. Compos. Sci.* **4** 6

Chapter 9

'I'

9.1 Interlaminar shear

Delamination is a detrimental failure mode in a composite and this is often related to the bond strength in shear between individual laminates. Characterising the interlaminar shear strength is therefore an important contribution to the design process of engineering composites and this section aims to provide some fundamental information on this property. Interlaminar shear strength tests are well documented and standardised. There are several approaches to the determination of interlaminar shear strength and these can be conducted via short beam tests (ASTM D2344, ISO 14130, EN 2377, EN 2563) [1–4], double beam shear tests (ISO 19927) [5], V-notched beam tests (ASTM D5379) [6], V-notched rail tests (ASTMD7078) [7], notched tensile tests (ASTM D3846, BS 4994) [8, 9], and notched compression tests (BS 6464) [10], though a few of these standards have been withdrawn. Of these, the short beam test method is very popular and this section will use the short beam test method as an exemplar. The short beam test method requires that the span-to-depth ratio of a composite sample is small. This is so that bending moments in a three-point bending test are effectively non-existent and lateral shear loading dominates the loading mode of the bending beam. ASTM D2344 for example recommends a specimen span-to-thickness ratio of 4, with a minimum thickness of 2 mm and a scheme of the testing set up is shown in figure 9.1. Using this method and assuming a unidirectional laminated composite with the fibre axis being in the x-direction, the interlaminar shear stress (ILSS) in a y-direction τ_{max} can be calculated according to equation (9.1), where P_{max} is the maximum loading, b is the specimen width and h is the specimen thickness.

$$\tau_{max} = \frac{3}{4}\frac{P_{max}}{wh} \tag{9.1}$$

doi:10.1088/978-0-7503-3527-0ch9

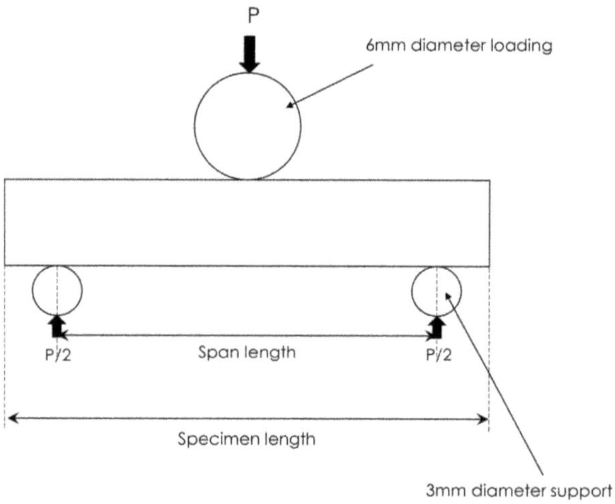

Figure 9.1. Horizontal shear loading set up in a short beam test.

9.2 Interphase

The interphase of a composite material is the region at the interface between reinforcement and matrix material where polymer molecules are pinned [11] and thus constrained to the surface of the reinforcement. The interphase has its own mechanical properties and behaviour, due to its constricted nature, however, these properties and behaviours are mostly evident at the lower length scales. Since this is a molecular level occurrence, the properties and behaviour of the interphase most noticeably contributes to the properties and behaviour of the composite at the molecular and nano-levels. Thus nanoparticle fillers with high ratios of surface area to volume will be particularly affected since nanoparticles are by definition particles with dimensions under 100 nm, and the interphase formed at interfaces in polymer matrix composites is typically between 1 and 50 nm in thickness [12]. This section will correlate the interphase to composite modulus and strength in nanocomposites.

9.2.1 Interphase influence on the Young's modulus

A popular model that can be used to predict the modulus of nanocomposites with a respectable level of accuracy against experimental data was proposed by Ji and co-workers in 2004 [13]. Their model considered particles as either spherical, layered or cylindrical and the ratio of the Young's modulus of the composite E_c to that of the matrix material E_m is shown in equation (9.2), where α and β are geometrically governed parameters relating to the interphase (equations (9.3), (9.4) and (9.5)) and the nanoparticle (equation (9.6)), respectively, and m is a ratio of the Young's modulus of the interphase, E_i, and the Young's modulus of the matrix, equation (9.7), and E_p is the Young's modulus of the nanoparticle. The calculation of α specifically for spherical, layered and cylindrical nanoparticles is shown in equations (9.3), (9.4) and (9.5). In these equations, φ_f is the volume fraction of the nanoparticle component of the composites, r is the nanoparticle radius, d is the nanoparticle thickness or diameter, and t is the thickness of the interphase around the nanoparticle.

$$\frac{E_c}{E_m} = \left(1 - \alpha + \frac{\alpha - \beta}{1 - \alpha + \frac{\alpha(m-1)}{\ln m}} + \frac{\beta}{1 - \alpha + \frac{(\alpha - \beta)(m+1)}{2} + \beta\frac{E_{\text{np}}}{E_m}}\right)^{-1} \tag{9.2}$$

$$\alpha_{\text{spherical}} = \sqrt{\left(\frac{t}{r} + 1\right)^3 \varphi_f} \tag{9.3}$$

$$\alpha_{\text{layered}} = \sqrt{\left(2\frac{t}{d} + 1\right) \varphi_f} \tag{9.4}$$

$$\alpha_{\text{cylindrical}} = \sqrt{\left(\frac{t}{r} + 1\right)^2 \varphi_f} \tag{9.5}$$

$$\beta = \sqrt{\varphi_f} \tag{9.6}$$

$$m = \frac{E_i}{E_m} \tag{9.7}$$

The interphase volume fractions, φ_i, for spherical, layered and cylindrical nanoparticles can subsequently be calculated using equations (9.8), (9.9), and (9.10), respectively.

$$\varphi_{i,\text{spherical}} = \varphi_f\left[\left(\frac{r+t}{r}\right)^3 - 1\right] \tag{9.8}$$

$$\varphi_{i,\text{layered}} = \varphi_f\left(\frac{2t}{d}\right) \tag{9.9}$$

$$\varphi_{i,\text{cylindrical}} = \varphi_f\left[\left(\frac{r+t}{r}\right)^2 - 1\right] \tag{9.10}$$

$$\alpha = \sqrt{\varphi_i + \varphi_f} \tag{9.11}$$

Coupling these equations (9.8), (9.10) and (9.10) with their equivalents with for α (equations (9.3), (9.4) and (9.5)), it can be seen that generally, α can be represented by equation (9.11), and this can be fed back into equation (9.2) for all types of nanocomposites [14].

9.2.2 Interphase influence on strength

The tensile strength of a nanocomposite will be affected by the presence of the interphase. Strength predictions are generally more complex and less accurate than stiffness predictions however a model developed by Pukansky [15] is often considered a very good baseline model as it fits well to experimental data from nanocomposites comprising a variety of differently shaped particles. While this model was not specifically developed for nanocomposites, its elements are entirely relevant to nanocomposites. The model is shown in equation (9.12), where σ_c and σ_m are the strengths of the composite and matrix, respectively, B is a parameter that reflects the interaction and adhesion between the reinforcement and matrix, and φ_f is the volume fraction of reinforcement.

$$\frac{\sigma_c}{\sigma_m} = \frac{1 - \varphi_f}{1 + 2.5\varphi_f} \exp\left(B\varphi_f\right) \tag{9.12}$$

$$B = \left(1 + A_{\text{np}}\rho_{\text{np}}t\right) \ln \frac{\sigma_i}{\sigma_m} \tag{9.13}$$

The empirical parameter B can be defined according to equation (9.13), where A_{np} is the surface area of the reinforcement, ρ_{np} is the density of the reinforcement, t is the thickness of the interphase, and σ_i is the strength of the interphase. Pukansky's fundamental model can be adapted to a wide variety of nanocomposites [16].

References

[1] ASTM D2344/D2344M-16 2016 *Standard Test Method for Short-Beam Strength of Polymer Matrix Composite Materials and Their Laminates* (West Conshohocken, PA: ASTM Int.)

[2] ISO 14130:1997 1997 Technical Committee: ISO/TC 61/SC 13 Composites and reinforcement fibresFibre-reinforced plastic composites—Determination of apparent interlaminar shear strength by short-beam method

[3] BS EN 2377:1989 *Pecification for glass fibre reinforced plastics. Test method. Determination of apparent interlaminar shear strength* (British Standards Int.)

[4] BS EN 2563:1997 *Carbon fibre reinforced plastics. Unidirectional laminates. Determination of the apparent interlaminar shear strength* (British Standards Int.)

[5] ISO 19927:2018 2018 *Fibre-reinforced plastic composites—Determination of interlaminar strength and modulus by double beam shear test* Technical Committee: ISO/TC 61/SC 13 Composites and reinforcement fibres

[6] ASTM D5379/D5379M-19e1 2019 *Standard Test Method for Shear Properties of Composite Materials by the V-Notched Beam Method* (West Conshohocken, PA: ASTM Int.)

[7] ASTM D7078/D7078M-20e1 2020 *Standard Test Method for Shear Properties of Composite Materials by V-Notched Rail Shear Method* (West Conshohocken, PA: ASTM Int.)

[8] ASTM D3846-08(2015) 2015 *Standard Test Method for In-Plane Shear Strength of Reinforced Plastics* (West Conshohocken, PA: ASTM Int.)

[9] BS 4994:1987 *Specification for Design and Construction of Vessels and Tanks in Reinforced Plastics* (British Standards Int.)

[10] BS 6464:1984 *Specification for Reinforced Plastics Pipes, Fittings and Joints for Process Plants* (British Standards Int.)
[11] Touaiti F, Alam P, Toivakka M and Bousfield D W 2010 Polymer chain pinning at interfaces in $CaCO_3$–SBR latex composites *Mater. Sci. Eng.* **A 527** 2363–9
[12] DeArmitt C 2017 Functional fillers for plastics *Applied Plastics Engineering Handbook* ed M Kutz and W Andrew 2nd edn (Elsevier: Oxford)
[13] Ji X L, Jing J K, Jiang W and Jiang B Z 2004 Tensile modulus of polymer nanocomposites *Polym. Eng. Sci.* **42** 983–93
[14] Zare Y and Rhee K Y 2020 A simple technique for calculation of an interphase parameter and interphase modulus for multilayered interphase region in polymer nanocomposites via modeling of Young's modulus *Phys. Mesomech.* **23** 332–9
[15] Pukansky B 1990 Influence of interface interaction on the ultimate tensile properties of polymer composites *Composites* **21** 255–62
[16] Zare Y and Rhee K Y 2020 A simple and sensible equation for interphase potency in carbon nanotubes (CNT) reinforced nanocomposites *J. Mater. Res. Technol.* **9** 6488–96

Chapter 10

'J'

10.1 Joining of metal matrix composites (MMCs)

Due to their unique properties, MMCs are gaining popularity in a number of areas including aerospace, automotives and in various marine technology sectors. The two primary methods for welding MMCs are fusion welding and solid-state welding. Fusion welding processes include gas tungsten arc welding, laser beam welding and electron welding, while solid-state welding includes friction welding and friction-stir welding. The processes are described individually below [1].

10.1.1 Fusion welding: gas tungsten arc welding (GTAW) or tungsten inert gas (TIG) welding

GTAW is an arc welding process where MMCS can be fused together by heating between a non-consumable tungsten electrode and a base metal. During heating, inert gas is discharged and allowed to flow around the electrode. While the process is excellent for joining metal alloys, the presence of ceramic particulates in MMCs causes problems. These problems stem from differences in the thermal expansion of reinforcing particulates and the metal matrix, the different melting temperatures of reinforcing particulates and the metal matrix, high arc energies leading to (ceramic) particulate degradation and weakened reinforcement–matrix interfaces, and surface reactions between the particulates and the metal matrix material. To circumvent some of these problems filler material can be added. The filler should ideally have high wetting properties to the reinforcing phases of the MMC whilst additionally having good affinity with the metal matrix.

10.1.2 Fusion welding: laser beam welding

Laser beam welding (LBW) involves the welding of MMCs at their interfaces using lasers to concentrate heat and melt material at the interface. Since the heat generated from the laser is highly focussed, the process can be used to weld together specific

doi:10.1088/978-0-7503-3527-0ch10

regions of surfaces (e.g. edges) without disrupting material in other regions of the same surface. Laser welding has an additional advantage over other fusion welding processes such as GTAW, as unlike GTAW, it does not need to be conducted in a vacuum. Since it is a focussed process able to create weld surfaces with high aspect ratios, the process has gained popularity in several industrial sectors including the aerospace and automotive industries.

10.1.3 Fusion welding: electron beam welding

Electron beam welding (EBW) involves the use of high energy electrons to generate heat at interfaces to create weldable surfaces. The process is conducted in a vacuum where the kinetic energy of the electrons is converted to heat energy on contact with the MMC surfaces. The surfaces melt and are conjoined at melted surface interfaces. On solidification, the interface becomes the weld. The process is popular in a very large number of industries.

10.1.4 Solid-state welding: friction welding

Friction welding (FW) involves heat generated at welding surfaces by friction (high speed rubbing). The MMC surfaces to be welded are typically the surfaces that are rubbed at high speed and the heat that develops locally, enables high pressure forging of the MMC parts. An advantage of FW is that it is a rapid process, and the mechanical properties of welded interfaces are generally quite predictable.

10.1.5 Solid-state welding: friction-stir welding

Friction-stir welding (FSW) involves heat generated at welding surfaces, similarly to FW, but rather than the friction coming from the rubbing of two MMC surfaces, it is created by a rotating pin tool. The rotating pin tool can be used to localise the areas to which friction is applied and the heat generated from friction softens the metal matrix in the region of the joining, and the joint is thus created in its solid state.

10.2 Joining

10.2.1 Butt joints

Butt joints are simple to manufacture but are mechanically inferior joints. Figure 10.1 shows examples of different fundamental butt joints including simple butt, scarf, double butt lap-joints and tongue and groove butt joints. Simple butt joints are often used to connect lines of composite piping or bars, however these joints are normally wrapped with wet-lay layers of composite to stabilise the otherwise weak joining. Scarf joints are often used to connect wood composites, as are double butt lap-joints, though these are essentially extended ply-drops in FRP composite structures such as wind and tidal turbine blades.

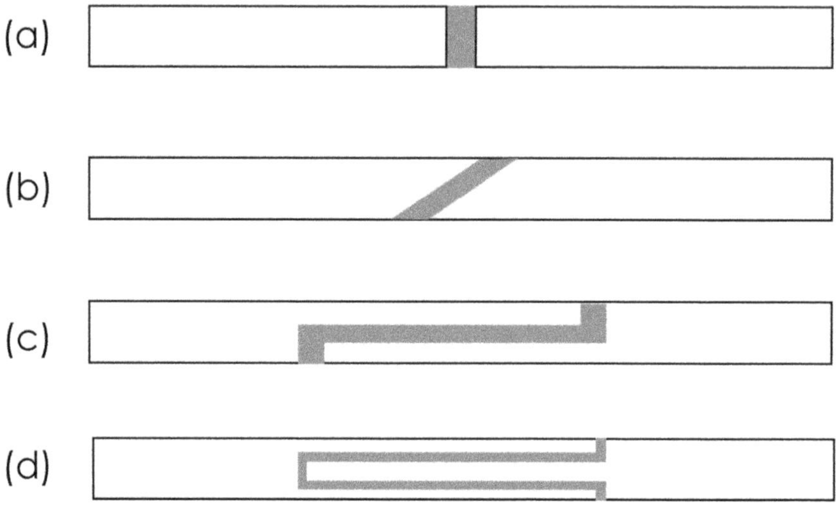

Figure 10.1. A few examples of butt joints: (a) simple butt joint, (b) scarf joint (also bevel joint), (c) soluble butt lap-joint and (d) tongue and groove butt joint.

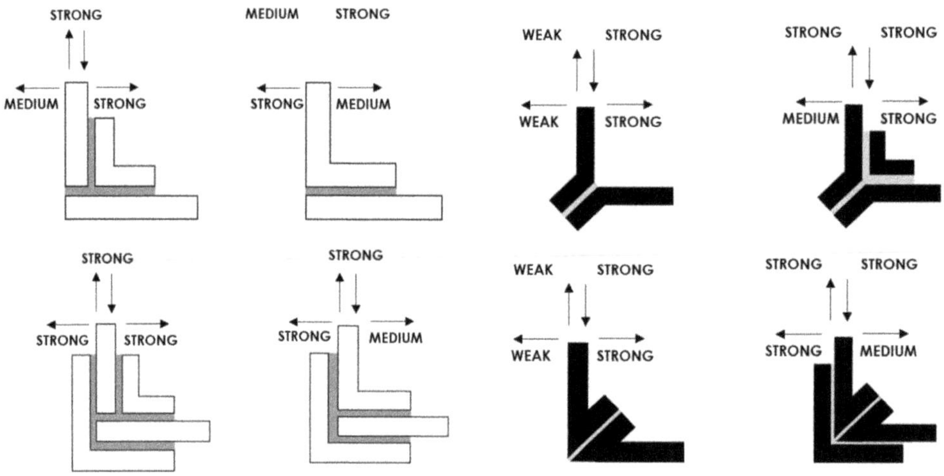

Figure 10.2. Examples of corner joints showing performance expectations with respect to loading direction (arrows).

10.2.2 Corner joints

Corner joints are most popular in structural composites (e.g. timber-based) as they are enablers for the modular construction of complex architectures. They also allow for the creation of tight corner angles, and as such are generating greater interest amongst fibre reinforced plastic (FRP)-based construction, where moulded corners are found to be insufficiently tight. Figure 10.2 provides examples of corner joints of interest to composites engineers, showing their strengths and weaknesses in particular directions of loading [2, 3].

10.2.3 Hybrid joints

Hybrid joints aim to improve joining between composite components, or composite materials, by making use of two or more joining mechanisms. Figure 10.3 shows a variety of different hybrid joining methods [3]. In figure 10.3(a) two or more types of adhesive can be used to improve joining. There are various reasons for this, some of which can be mechanical (e.g. mixed-mode loading) and others which can be practical (e.g. water-proofing). In figures 10.3(b) and (c) mechanical connectors such as bolts or rivets are shown as being used in conjunction with adhesives, and in figure 10.3(d) composite plates (e.g. MMCs) are spot welded and this is combined with a structural adhesive to finish the joining.

10.2.4 Lap-joints

Lap-joints are amongst the most common composite joint types. Figure 10.4 provides a few variations of the lap-joint. Single shear lap-joints (a) are easy to manufacture but suffer from peeling-type failures due to the misalignment of laminates. This problem can be circumvented to some extent using joggle lap-joints (b) as the primary body of composite material remain aligned, improving the regularity of the global bending moment, and decreasing stress concentrations at the corners of the lap. If reduced bending moments are desired, then double lap-joints (c) are a simple alternative, though these can be difficult to fabricate.

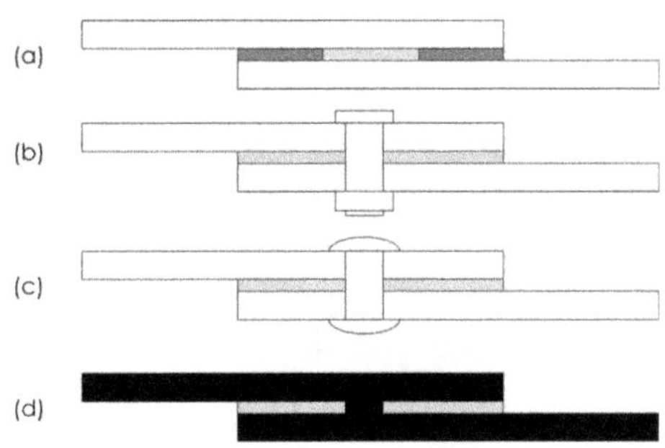

Figure 10.3. Examples of hybrid methods of joining: (a) mixed adhesive joining, (b) bolt-adhesive joining, (c) rivet-adhesive joining and (d) weld-adhesive joining.

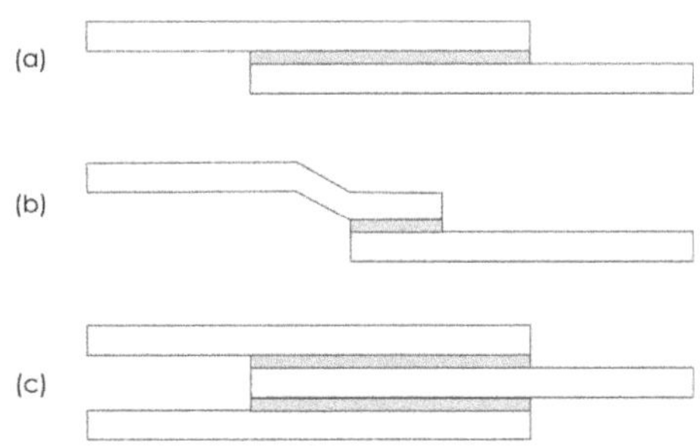

Figure 10.4. Examples of lap-joints: (a) single lap-joint, (b) joggle lap-joint and (c) double lap-joint.

10.2.5 Strap joints

Strap joints, figure 10.5, behave similarly to lap-joints in that the single strap joints (a) experience high stress concentrations at bonded corners and thus have a tendency of peeling, while double strap joints (b) have high bending moments and are geometrically symmetrical, and are thus both strong as well as dimensionally stable.

10.2.6 T-joints

T-joints are similarly to corner joints, most popular in structural composites (e.g. timber-based) as they are enablers for the modular construction of complex architectures. Figure 10.6 provides examples of T-joints of interest to composites engineers, showing their strengths and weaknesses in particular directions of loading [2, 3].

10.3 Joining of thermoplastic matrix composites

Thermoplastic matrix composites are receiving more attention as they are recognised as high performance materials with an option of post-use melting and recycling that thermoset matrix composites are unable to offer. Since the thermoplastic matrix

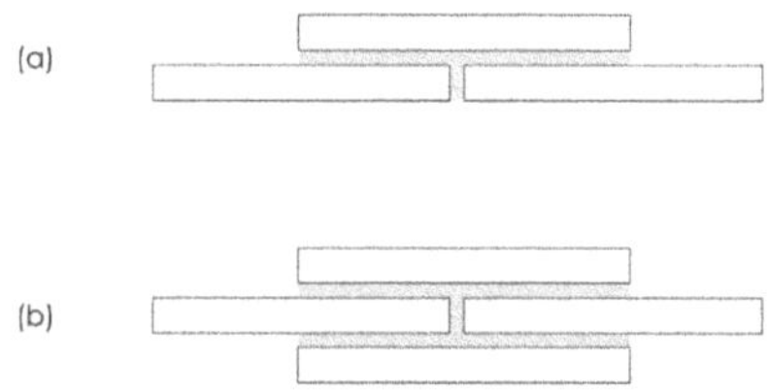

Figure 10.5. Examples of strap joints: (a) single strap joint and (b) double strap joint.

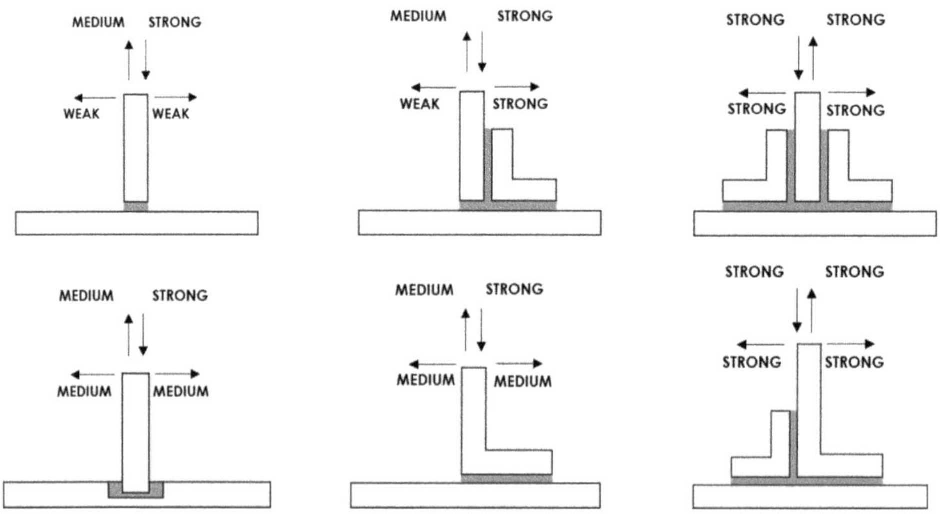

Figure 10.6. Examples of T-joints showing performance expectations with respect to loading direction (arrows).

of these composites softens and melts under heat, there are additional benefits in joining composite parts through welding processes. Fusion welding processes show great promise for the joining of thermoplastic matrix composites, in particular: ultrasonic welding, induction welding and resistance welding. This section will introduce each of these fusion welding processes.

10.3.1 Ultrasonic welding

In this welding process, two thermoplastic matrix composite parts are held together under pressure and subjected to ultrasonic vibrations up to 50 Hz, normal to the contact surface. Figure 10.7 provides schematic representation of the process [4]. Sinusoidal strains caused by both high and low amplitude oscillations gives rise to intermolecular polymer chain motion, resulting in friction and consequently the production of heat. The main parameters by which means the process is controlled include the pressure applied normal to the contact surface, the time of exposure to ultrasonic vibration, the amplitude and frequency of the vibration, and the solidification time (or hold time).

10.3.2 Induction welding

Induction welding involves the heating of thermoplastic matrix composite contact surfaces with an electrically conducting element. Power must be supplied to enable induction heating, which takes place by electromagnetic induction where heat is generated by eddy currents. Heat generation is affected by certain physical parameters of the composite, including level of consolidation, fibre architecture, the lay-up of the composite and the properties of the polymer matrix. The main parameters by which means the process is controlled include the frequency, the input power, the time of welding exposure and the applied contact pressure. Figure 10.8 provides a schematic representation of the process [4].

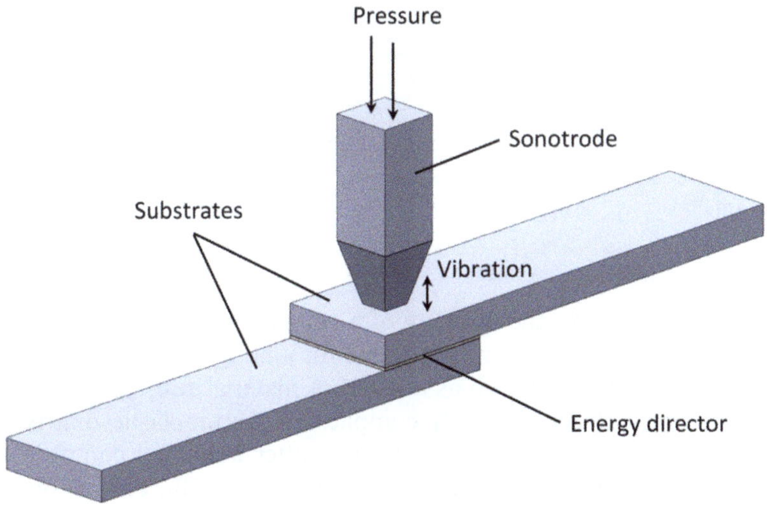

Figure 10.7. Schematic representation of ultrasonic welding process. Reprinted from [4] with permission from MDPI (CC BY 4.0).

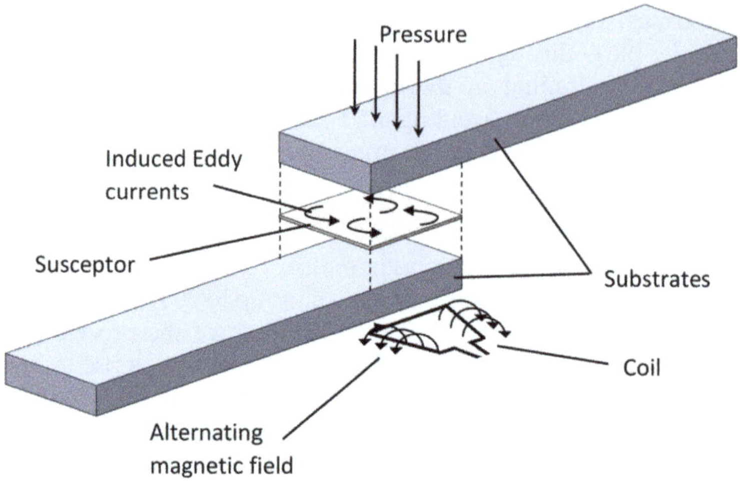

Figure 10.8. Schematic representation of induction welding process. Reprinted from [4] with permission from MDPI (CC BY 4.0).

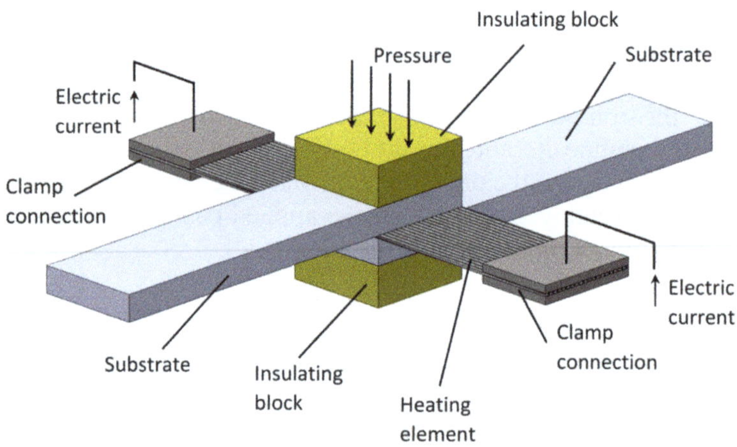

Figure 10.9. Schematic representation of resistance welding process. Reprinted from [4] with permission from MDPI (CC BY 4.0).

10.3.3 Resistance welding

Resistance welding involves the trapping of a conductive implant at the interfaces of thermoplastic matrix composite parts intended for joining. Electrical currents supplied to the implant causes it to heat up as a result of material resistance. This method is good for the joining of geometrically complicated thermoplastic matrix composite materials as the implant remains within the joint after welding is complete. The main parameters by which means the process is controlled include the contact pressure at the welding interfaces, the time of exposure to welding, and the welding current. Figure 10.9 provides a schematic representation of the process [4].

References

[1] Kalaiselvan K, Dinaharan I and Murugan N 2020 Routes for the joining of metal matrix composite materials *Encyclopedia of Materials: Composites* (Amsterdam: Elsevier)
[2] Adams R D, Comyn J and Wake W C 1997 *Structural Adhesive Joints in Engineering* 2nd edn (London: Chapman and Hall)
[3] da Silva L F M 2011 Design rules and methods to improve joint strength *Handbook of Adhesion Technology* ed L F M da Silva, A Ochsner and R D Adams (Berlin: Springer)
[4] Reis J P, de Moura M and Samborski S 2020 Thermoplastic composites and their promising applications in joining and repair composites structures: a review *Materials* **13** 5832

Chapter 11

'K'

11.1 Kevlar

Kevlar fibres belong to the broad group known as aramids, or aramid fibres. They were originally developed by Du Pont in the 1970s and are long chain synthetic polyamides. At the molecular level, 85% or more, of the amide groups (-CO-NH-) in Kevlar are directly connected to two aromatic rings. The structure is formed by reacting 1,4-phenylene-diamine with terephthaloyl chloride and its molecular chains are aligned through the process of drawing. Since Kevlar is a liquid-crystalline polymer, when drawn, its molecular chains will align under intermolecular shear to the direction of draw. An important consequence of aligning the molecular chains in this way is that the (-NH-) ends of the amide groups on one molecule will form hydrogen bonds with the (-CO-) ends of amide groups on neighbouring molecules. This results in the creation of vast numbers of hydrogen bonds linking Kevlar molecules together, figure 11.1, and the fibres made up of these tightly bonded molecules realise high mechanical properties such as tensile strength and stiffness.

Kevlar is produced in various forms including continuous fibres (e.g. Kevlar-29 and Kevlar-49), as Kevlar pulp, Kevlar staple, and Kevlar floc. Table 11.1 [1], compares a few properties of Kevlar-29 and Kevlar-49, note that for epoxy resin impregnated fibres (ASTM D2343 [2]), the tensile strengths of Kevlar-29 and Kevlar-49 are the same at 3.62 GPa. The specific heat capacity (at 25°), thermal conductivity and shrinkage of each of Kevlar-29 and Kevlar-49 are invariably the same at 1.42 J (g °C)$^{-1}$, W (m K)$^{-1}$ and < 0.1%, respectively.

Kevlar yarns are often woven into textiles and are used in a broad range of applications including aerospace engineering, body armour and bullet proof vests, as elastomer stiffening agents (as floc), in car brake pads (as pulp), and in the fuel tanks of racing cars. Aside from its obvious mechanical benefits, it also has short-term degradation resistance to a number of corrosive compounds. While immersion in high concentration hydrochloric acid in 24 h, for example, has no effect on the mechanical

doi:10.1088/978-0-7503-3527-0ch11

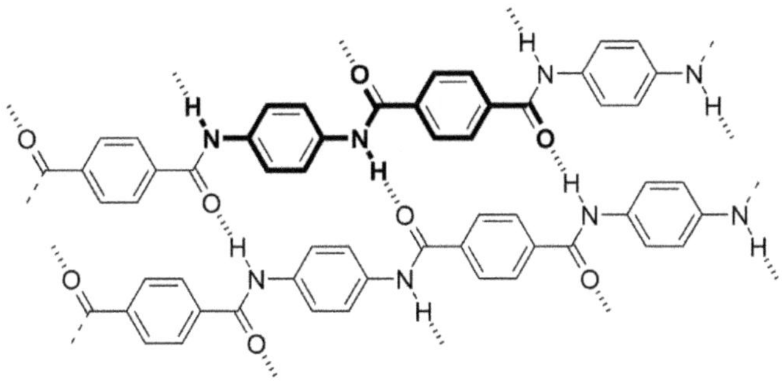

Figure 11.1. The molecular structure of drawn Kevlar showing hydrogen bonding between the (-NH-) and (-CO-) of neighbouring amide groups and a monomer unit (in bold). Wikipedia: Hbf878 (CC0 1.0).

Table 11.1. Properties of Kevlar-29 and Kevlar-49.

Material	Density ($g\ cm^{-3}$)	Tensile strength (GPa)	Modulus (GPa)	Elongation (%)	Decomposition °C
Kevlar-29	1.44	2.92	70.3	3.6	427–482
Kevlar-49	1.44	3.00	112	2.4	427–482

strength of Kevlar fibres, immersion of fibres for longer periods of time (even in lower concentrations of hydrochloric acid) results in significant levels of degradation.

11.2 Knitted fabric composites

Knitted fabric composites fit into the larger family of continuous fibre reinforced plastics (FRPs). They typically have medial in-plane strength (bi-axial) and stiffness properties and this is partly due to that the continuous fibres have different angles or orientation, and partly due to waviness resulting from the interlinking of warp and weft fibres. One of the larger benefits of knitted fabric composites is their ease of processing, even when using stiff high performance engineering fibres such carbon and glass. Warp and weft knitting are different knitting techniques that result in different fabric types. In weft knitting, there is one set of yarn used in the weft-wise direction, while in warp knitting, many sets of yarns are used in the warp-wise direction. Knitted fibre composites are commonly knitted into preform fabrics using a warp knit as it is known to be a good knit for load-bearing [3]. This said, weft knitted fabrics are stretchy and are thus formed more easily than warp knits into a variety of complex shapes [4]. Essentially, warp knits are dimensionally stable (similar to a woven) while weft knits are low-stability, high-stretch fabrics. Either thermoset or thermoplastic matrices are used as the matrix for the fabric, which is often sheet formed or draped (over a mould). Knitted fabrics differ from woven textiles and from non-wovens in a variety of ways. While woven textiles are linearly overlaid/orthogonally interlaced materials, and non-wovens are essentially randomly distributed short fibres, knitted fabrics are comprised of systematically arranged looped, linked structures, figure 11.2 [5]. The looping and linking creates a level of geometrical complexity that influences not only the mechanical behaviour of the textile, but also its resistance to fibre-fibre separation (which can be measured as tear). A few examples of different knits are shown in figure 11.3 [6].

Since knitted fabrics are geometrically rather complex structures, it can be hard to predict their mechanical properties such as strength and modulus. Modulus

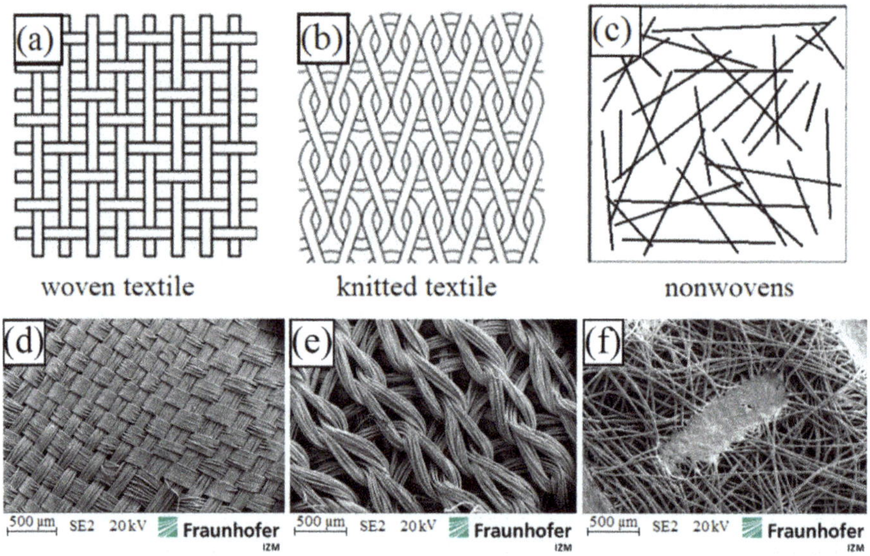

Figure 11.2. Architectural differences between woven textiles, knitted fabrics and non-wovens: schematics of (a) woven; (b) knitted and (c) nonwoven fabrics and SEM images below of: (d) woven; (e) knitted and (f) nonwoven fabrics. Reprinted from [5] with permission from MDPI (CC BY 4.0).

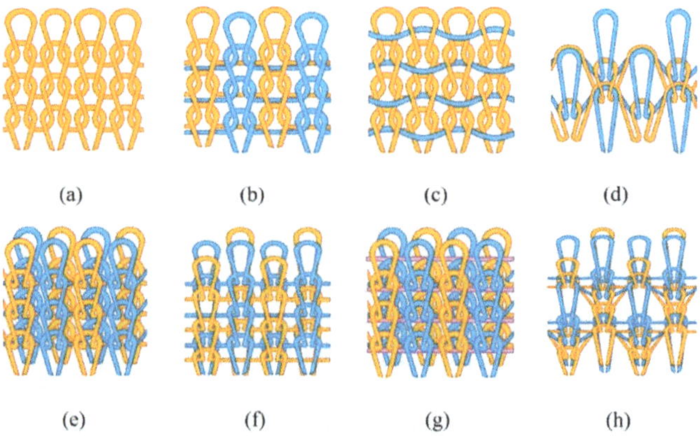

Figure 11.3. Different knits: different structures: (a) plain knit, (b) 1 ×1 plain knit, (c) plain knit with weft-inlay, (d) double pique, (e) interlock, (f) variable half milano, (g) interlock with weft-inlay and (h) variable half cardigan. Reprinted from [6] with permission from MDPI (CC BY 4.0).

predictions are generally simpler than strength predictions and a few models have been shown to be good at predicting the in-plane Young's moduli (E_{11} and E_{22}), showing reasonable agreement with experimental values. One example is shown in equation (11.1), where E_c is the Young's modulus of the knitted fabric composite, E_m is the Young's modulus of the matrix polymer, E_f is the axial Young's modulus of fibre reinforcement, V_m is the volume fraction of the matrix, V_f is the volume fraction of fibre reinforcement, and η is known as the efficiency of a curved fibre in its ability to bear load and is represented in equation (11.2), where $a\theta$ is the proportion of fibre oriented at an angle of θ and l is the length of the fibre [7].

$$E_c = E_m V_m + E_f V_f \eta \tag{11.1}$$

$$\eta = a\theta \cos^4 \theta dl \tag{11.2}$$

Apart from benefits arising from the ease of processing, another reason for why knitted fabric composites are used is for their energy absorbing properties. Crack resistance in knitted fabric composites is achieved through the complex geometry of the knit. The looped structures of the reinforcing fibres force cracks to branch out into micro-cracks, which retards crack growth and leads to higher toughness. As a general rule, smaller knitted loops will result in increased crack branching and thus improved fracture toughness. When comparing the fracture toughness of E-glass knitted fabric reinforced epoxy against unidirectional (UD) E-glass reinforced epoxy, the former can be expected to have values in the region of 1200 J m^{-2}, while the latter will have values in the region of 5–700 J m^{-2} [7]. Unlike UD-FRP or woven composites therefore, knitted fabrics, due to the looped and linked reinforcing, do not fracture abruptly during loading but tend to carry some load after first crack. Thus, although the in-plane mechanical properties of knitted fabric composites are inferior to those of UD-FRP, the fracture toughness properties are superior.

References

[1] 2019 *Kevlar Aramid Fibre Technical Guide* (Wilmington, DE)

[2] ASTM D2343-17 2017 *Standard Test Method for Tensile Properties of Glass Fiber Strands Yarns, and Rovings Used in Reinforced Plastics* (West Conshohocken, PA: ASTM International)

[3] Au K F 2011 *Advances in Knitting Technology* (Woodhead: Cambridge)

[4] Pham M Q, Dobrich O, Trumper W, Gereke T and Cherif C 2019 Numerical modelling of the mechanical behaviour of biaxial weft-knitted fabrics on different length scales *Materials* **12** 3693

[5] Dils C, Werft L, Walter H, Zwanzig M, Krshiwoblozki Mv and Schneider-Ramelow M 2019 Investigation of the mechanical and electrical properties of elastic textile/polymer composites for stretchable electronics at quasi-static or cyclic mechanical loads *Materials* **12** 3599

[6] Siyao M, Liu S, Peihua Z and Hairu L 2020 Functional investigation on automotive interior materials based on variable knitted structural parameters *Polymers* **12** 2455

[7] Kelly A and Zweben C 2000 *Comprehensive Composite Materials* (Pergamon: Oxforfd)

IOP Publishing

Composites Engineering: An A–Z Guide

Parvez Alam

Chapter 12

'L'

12.1 Laminate theory

In laminate theory (LT), also called composite laminate theory (CLT), ply stiffness can be assumed to be under a state of plane stress, equation (12.1), where Q_{ij} is termed the reduced stiffness, and σ_1 and σ_2 are normal stresses, σ_6 is shear stress, ϵ_1 and ϵ_2 are normal strains, and ϵ_6 is shear strain. Q_{ij} is related to stiffness, C_{ij} as $Q_{ij} = C_{ij} - \frac{C_{i3}C_{j3}}{C_{33}}(i, j = 1, 2, 6)$. σ for a 0° lamina can be calculated thence as $\sigma_i = Q_{ij}\epsilon_j$.

$$\begin{Bmatrix} \sigma_1 \\ \sigma_2 \\ \sigma_6 \end{Bmatrix} = \begin{bmatrix} Q_{11} & Q_{12} & 0 \\ Q_{21} & Q_{22} & 0 \\ 0 & 0 & Q_{33} \end{bmatrix} \begin{Bmatrix} \epsilon_1 \\ \epsilon_2 \\ \epsilon_6 \end{Bmatrix} \tag{12.1}$$

Q_{ij} relates to engineering constants as: $Q_{11} = \frac{E_{11}}{1 - \nu_{12}\nu_{21}}$; $Q_{12} = \nu_{21}Q_{11}$; $Q_{22} = \frac{E_{22}}{1 - \nu_{12}\nu_{21}}$; $Q_{16} = 0$; $Q_{26} = 0$; $Q_{66} - G_{12}$, where E is elastic modulus, ν is Poisson's ratio, and G is shear modulus.

Ply angles in laminates are often varied. Calculating the reduced stiffness so as to take into account the ply angle, $\bar{Q}_{ij}$, requires consideration of the effects of these angles on the properties, equation (12.2), where $m = \cos(\theta)$ and $n = \sin(\theta)$, θ being the ply orientation angle.

$$\begin{Bmatrix} \bar{Q}_{11} \\ \bar{Q}_{22} \\ \bar{Q}_{12} \\ \bar{Q}_{66} \\ \bar{Q}_{16} \\ \bar{Q}_{26} \end{Bmatrix} = \begin{bmatrix} m^4 & n^4 & 2m^2n^2 & 4m^2n^2 \\ n^4 & m^4 & 2m^2n^2 & 4m^2n^2 \\ m^2n^2 & m^2n^2 & m^4 + n^4 & -4m^2n^2 \\ m^2n^2 & m^2n^2 & -2m^2n^2 & (m^2 - n^2)^2 \\ mn^3 & -mn^3 & n^3m - m^3n & 2(n^3m - m^3n) \\ mn^3 & -m^3n & m^3n - n^3m & 2(m^3n - n^3m) \end{bmatrix} \begin{Bmatrix} Q_{11} \\ Q_{22} \\ Q_{12} \\ Q_{66} \end{Bmatrix} \tag{12.2}$$

$\bar{Q}_{ij}$ can then be used to determine the extensional stiffness (A_{ij}), equation (12.3), the coupling stiffness (B_{ij}), equation (12.5), and the bending stiffness (D_{ij}), equation (12.4), where t_k is the thickness of the kth lamina and $\bar{z}_k$ is the distance from the mid-plan to the centroid of the kth lamina.

$$A_{ij} = \sum_{k=1}^{n} \left(\bar{Q}_{ij}\right)_k t_k \tag{12.3}$$

$$B_{ij} = \frac{1}{2} \sum_{k=1}^{n} \left(\bar{Q}_{ij}\right)_k t_k \bar{z}_k \tag{12.4}$$

$$D_{ij} = \frac{1}{3} \sum_{k=1}^{n} \left(\bar{Q}_{ij}\right)_k \left(t_k \bar{z}_k^{-2} + \frac{t_k^3}{12} \right) \tag{12.5}$$

These in turn can then be used to calculate the forces, N, and moments, M, according to equation (12.6), which is shown in the expanded form in equation (12.7)

$$\begin{Bmatrix} N \\ M \end{Bmatrix} = \begin{bmatrix} A & B \\ B & D \end{bmatrix} \begin{Bmatrix} \epsilon^0 \\ \kappa \end{Bmatrix} \tag{12.6}$$

$$\begin{Bmatrix} N_{xx} \\ N_{yy} \\ N_{xy} \\ M_{xx} \\ M_{yy} \\ M_{xy} \end{Bmatrix} = \begin{bmatrix} A_{11} & A_{12} & A_{16} & B_{11} & B_{12} & B_{16} \\ A_{12} & A_{22} & A_{26} & B_{12} & B_{22} & B_{26} \\ A_{16} & A_{26} & A_{66} & B_{16} & B_{26} & B_{66} \\ B_{11} & B_{12} & B_{16} & D_{11} & D_{12} & D_{16} \\ B_{12} & B_{22} & B_{26} & D_{12} & D_{22} & D_{26} \\ B_{16} & B_{26} & B_{66} & D_{16} & D_{26} & D_{66} \end{bmatrix} \begin{Bmatrix} \epsilon_{xx}^0 \\ \epsilon_{yy}^0 \\ \epsilon_{xy}^0 \\ \kappa_{xx} \\ \kappa_{yy} \\ \kappa_{xy} \end{Bmatrix} \tag{12.7}$$

Here, κ_x, κ_y, and κ_{xy} are the laminate curvatures calculated as $\frac{1}{r}$, where r is the radius of curvature.

IOP Publishing

Composites Engineering: An A–Z Guide

Parvez Alam

Chapter 13

'M'

13.1 Mat

13.1.1 Structural/protective mats

Composite mats are sheets of fibrous material connected together using binders at the fibre–fibre interfaces. They are therefore, non-woven fibrous sheets. Composite mats are made using a range of materials including wood-based fibres, other natural fibres, and synthetic polymers such as high density polyethylene (HDPE). HDPE mats have certain advantages over wood-based mats such as superior weathering and lower densities. Both wood and HDPE mats are generally used for heavy duty outdoor activities where soil protection and load bearing are of primary importance.

13.1.2 Chopped strand mats (CSM)

Chopped glass fibre mats, or chopped strand mats (CSM) are utilised in the manufacture of large scale glass fibre reinforced plastic (FRP) moulds (e.g. boat hulls). CSM is most commonly used with polyester resins and enable the manufacture of thick moulded composite parts. They may also be used in conjunction with (e.g. in between) glass fibre fabrics/wovens for large composite moulds. These mats are manufactured by chopping short lengths of roving from continuous strands and are dispersed randomly over a moving belt to create a network of short strands and fibres. Styrene binder is then applied to the loose strands of material to form a CSM sheet. Alongside polyester resins, vinyl ester resins may also be used as when either of these wet-out the CSM, the styrene binder is dissolved and the mat is left flexible, conforming well to complex moulds. Though some do use epoxy resin to wet-out CSM, unlike polyester or vinyl ester, epoxy does not dissolve the styrene and this can leave the mat stiff, compromising therefore its ability to conform well to complex moulds. CSM is considerably less stiff and is weaker than woven or fabric glass fibre sheet. The architectural difference between CSM and cross-weave glass roving is shown in figure 13.1.

doi:10.1088/978-0-7503-3527-0ch13

Figure 13.1. Visible architectural differences between chopped strand mat (CSM) and a cross-weave glass roving. Reprinted from [1] with permission from MDPI (CC BY 4.0).

Table 13.1. Examples of materials used in biomedical composite mats, their means of manufacture, and their applications.

Materials	Manufacture	Application	Source
Graphene oxide/titanium dioxide/silk fibroin	Electrospinning	Scaffolds for tissue engineering	[2]
Chitosan/polydopamine/silk fibroin	Layer-by-layer assembly	Antibacterial performance	[3]
Poly (hydroxybutylate-co-hydroxyvalerate) (PHBV)/modified keratin	Electrospinning	Fibroblast growth	[4]
Hyroxyapatite/titanium dioxide/collagen type I	Electrospinning	Osteoblastic growth (bone scaffold)	[5]
Bacterial cellulose/kaolin	Pressure moulding	Wound healing	[6, 7]
Carboxymethyl chitosan/polyethylene oxide	Centrifugal spinning	Wound dressing	[8]

13.1.3 Biomedical composite mats

Composite mats are additionally prevalent in the biomedical sectors as they have a myriad of uses including as scaffolds for tissue repair and regeneration, for drug release, and as antibacterial sheets. Mats are made from a wide variety of materials and the main manufacturing methods range from simple compacting methods to more complex electrospinning approaches. They can be laminated to form part of multifunctional layered composite structures. Table 13.1 provides examples of materials used to make biomedical composite mats, their manufacturing methods and their applications.

13.2 Matrix

A matrix is a material, most often homogeneous and isotropic, that binds together the reinforcement phase of a composites. The matrix provides overall form to the composite, and enables load transfer between the reinforcement phases. There are three main classes of material that are used as matrix and these include polymers, ceramics and metals. A fourth material included in this section for completeness is carbon, which is used as a matrix material in carbon–carbon composites.

13.2.1 Carbon matrix

Carbon exists in a variety of different forms, e.g. graphite, diamond, amorphous carbon, single and multiwalled carbon nanotubes (CNT) and others. Carbon fibre/carbon matrix (C/C) composites are a special breed of composite primarily devoted to high temperature applications (1000–2200 °C) where strength and stiffness via a carbon reinforcement are required and the reinforcing matrix is a pyrolysed or chemically deposited carbonaceous material. Carbon oxidation at lower temperatures (ca. 450 °C) can be achieved using diffusion barriers or inhibitors to slow down the rate of reaction between oxygen and carbon. Chawla [9] describes three common processing techniques for C/C composites and these are described below:

- **Thermoplastic impregnation pyrolysis:** In this method, a preform (such as a carbon fibre woven) is impregnated by a thermoplastic pitch under the conditions of heat and pressure. The thermoplastic pitch is pyrolysed into carbon and the process is repeated until a desired carbon matrix density is achieved.
- **Thermoset impregnation pyrolysis:** In this method, a carbon fibre thermoset plastic matrix composite is manufactured using a conventional manufacturing method. The thermoset resin matrix is then pyrolysed into a carbonaceous matrix and the process is repeated until a desired carbon matrix density is achieved.
- **Chemical vapour deposition:** In this method, carbon is deposited onto and between carbon fibre reinforcements from a gaseous phase. Methane is an example precursor for the process.

13.2.2 Ceramic matrix

Ceramic matrix composites (CMCs) are applied where thermal protection coupled to mechanical performance is desirable. Examples include gas turbine engines where the operating temperatures may reach over 1000 °C–1300 °C, thermo-mechanical protection during re-entry (ca. 1500 °C) in the aerospace sectors, and heat exchangers in the energy sector. CMCs tend to have excellent properties of strength and stiffness when compared against unreinforced ceramics, and have improved properties of fracture toughness as a result of the embedded reinforcements. Importantly, since ceramics have very high melting and decomposition temperatures, and their atomic structures remain stable at elevated temperatures, CMCs are able to maintain excellent mechanical performance even at such elevated temperatures. Other major benefits in using ceramic matrices include high hardness, superior wear resistance, and superior corrosion resistance, each of which also remains at elevated temperatures. There are various generic manufacturing routes for incorporating the ceramic matrix phase into a CMC and these are listed below [10]:

- **Liquid phase processing:** There are three main liquid phase processing routes that enable the addition of a ceramic matrix to a reinforcing phase. These include reactive melt infiltration (RMI), polymer impregnation and pyrolysis (PIP), and slurry impregnation and hot processing (SIHP).
- **Vapour phase processing:** Vapour phase processing are chemical vapour infiltration (CVI) routes that enable the addition of a ceramic matrix to a reinforcing phase. CVI essentially involves the deposition of interphase materials and ceramic matrix to a reinforcing preform. There are four main types of CVI and these include isothermal/isobaric CVI (I-CVI), forced (temperature and pressure gradient) CVI (F-CVI), pressure-pulsed CVI (P-CVI) and film boiling.
- **Solid state processing—dispersion strengthening:** Dispersed particle, short fibre or whisker phases can be added to glass or ceramic matrices to improve the properties and functionality of the matrix. Dispersion strengthening is a process conducted under heat and as such, differences in the thermal expansion of reinforcement and matrix materials can affect the local stresses (both radial, σ_r, and tangential, σ_t) that develop in the matrix and may cause weakening of the matrix such that $-\sigma_r = 2\sigma_t = \frac{\Delta k \Delta T}{(1 + \nu_m / 2E_m) + (1 - 2\nu_p / E_p)}$, where k is the coefficient of thermal expansion, T is temperature, ν_m and ν_p are the Poisson's ratios of matrix and particle, respectively, and E_m and E_p are the Young's moduli of matrix and particle, respectively [11].

Chawla [12] lists the following as ideal attributes of ceramic matrix materials:

- It should be able to infiltrate bundles of fibres, particles, whiskers or a preform.
- It should form a mechanical or frictional bond with the reinforcement.
- It should not react chemically with the reinforcement during processing, or through its service-life.
- It should not exert physical damage to the fibre (e.g. through thermal expansion or other).
- It should have good resistance to mechanical loading such as creep, fatigue and impact, and should have good properties of toughness.
- It should be chemically stable, impermeable to moisture, non-oxidative, non-volatile.

Some common and important engineering ceramic matrix composite materials include alumina, silicon nitride, silicon carbide, silica, mullite (alumina-silica), boron nitride, boron carbide, and sialons (solid solution of alumina in silicon nitride).

13.2.3 Metal matrix

The largest current market for metal matrix composites (MMCs) is the automotive industry. MMCs in the automotives are valued for their contributions to lighter, high-performance materials used in, e.g. braking systems, cylinder heads, engine blocks transmission housings, valve covers, suspension links, door frames, crankshafts, piston/piston pins, and more. They are also of importance in the aerospace industry as they have the desirable properties of excellent strength and stiffness under elevated temperatures, and low densities in comparison to steel and nickel alloys [13]. F-119 and F-22 fighter jets for example include piston rods that are made up of titanium matrix MMCs [14].

Aluminium/aluminium alloy is by far the most commonly used metal matrix, though other metals and alloys are also occasionally used such as magnesium/magnesium alloy, titanium/titanium alloy, copper/copper alloy, and super metal alloys. Metal matrices such as these are often combined with ceramic reinforcements such as silicon carbide, zirconia, alumina, titanium carbide, boron carbide and sometimes carbon-based reinforcements such as graphite or nanotubes. Some benefits of the different metal matrices in MMCs mentioned here, are summarised below:

- **Aluminium matrix:** Some of the benefits of using aluminium as a matrix in MMCs when compared against pure aluminium alloy include high specific properties (strength and stiffness), improved creep resistance even under elevated temperatures, and increased resistance to wear. The disadvantages include reduced toughness, reduced ductility and increased complexity of the manufacturing process. While aluminium matrix MMCs are lower cost than CMCs, they have lower maximum service temperatures at which they can be used.
- **Magnesium matrix:** Magnesium matrix MMCs have similar benefits to aluminium MMCs. They are nevertheless more difficult to manufacture and have lower thermal conductivities, which limit their breadth of use. The lower density of magnesium alloys as compared to aluminium, have still made them attractive materials in the aerospace sectors where weight savings are of importance.
- **Titanium matrix:** Titanium matrix MMCs (fibre reinforced) are used in aircraft structures and missiles over aluminium matrix MMCs as their alloys are superior at retaining strength at elevated temperatures. Titanium alloys used as matrix in MMCs have the additional advantage of high resistance to corrosion. The primary difficulty with using titanium alloy as a matrix, is that it is very reactive and may react with the reinforcement during processing, to the detriment also of the interface.
- **Copper matrix:** Copper matrix MMCs are excellent candidate matrix materials when both thermal conductivity as well as the retention of strength under elevated temperatures are required.
- **Super alloy matrix:** Super alloys are nickel, iron-nickel or cobalt alloys. These are high-performance alloys that are still functional normally even when the temperature is elevated to close to the super alloy melting point. Super alloy MMCs are therefore attractive as they have excellent resistance to creep even

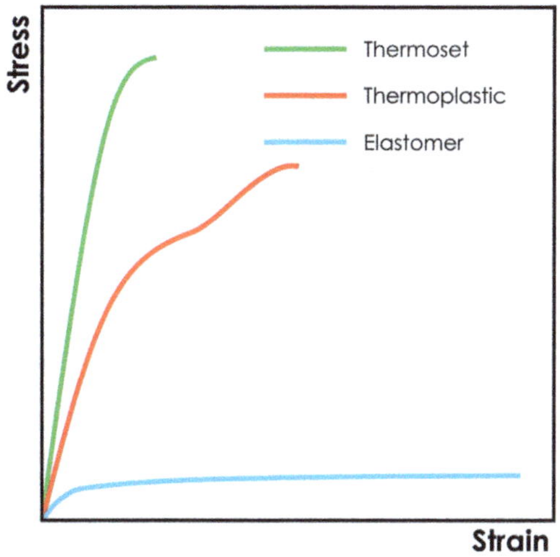

Figure 13.2. Generic comparative examples of stress–strain curves for thermosetting plastics, thermoplastics, and elastomers.

when exposed to prolonged elevated temperatures, have superior mechanical properties, and outstanding resistance to corrosion.

There are four main methods by which means MMCs can be manufactured and these include diffusion bonding under high pressures, casting, powder-metallurgy and physical vapour deposition methods.

13.2.4 Polymer matrix

Polymer matrix composites (PMCs) are extremely versatile in terms of design and manufacture, and consequently have a broad range of end applications. The choice of matrix will depend on a variety of factors including cost, availability, polymer chemistry affecting the characteristics of adhesion, and polymer chemistry affecting the bulk polymer properties (e.g. mechanical, optical, insulation, physical, hygroscopic). Polymer matrices can be subdivided into three groups: thermoplastics, thermosets and elastomers. While elastomers are usually thermosets (and require vulcanisation), they can also be thermoplastics and as such are placed into their own group of matrix material. The mechanical properties of each of these groups are rather diverse but in general, these can be summarised as shown in figure 13.2. These properties are ultimately related to the general molecular structures making up the specific polymer group. Thermosets are highly cross-linked rigid 3D-networked structures at the molecular level, and this prevents sliding of polymers. The result is that energy is stored within the material until fracture, the polymers are usually stiff and strong, and the stress–strain curves are close to linear to failure below the glass transition temperature (Tg). Thermoplastics generally do not exhibit much intermolecular cross-linking and their flexible molecules slide under loading, which gives the polymer a degree of toughness that thermosets do not have. As such, thermoplastics generally tend to be lower in stiffness and strength, but absorb more mechanical energy on loading than thermosets. Due to the alignment of polymer molecules, secondary force interactions often develop in thermoplastics and they strain harden. Elastomers are partly cross-linked between linear molecular chains and their molecules are somewhat folded and concertina-like in a contracted state. As such, on loading their molecules will elongate and will have a large elongation range for recovery, which is partially aided by the presence of occasional cross-links.

Generally, thermoplastics are more expensive than thermosets and as such have a lower cost-performance index. Thermoplastics are additionally soluble in organic solvents, can be softened through heating (thermosets decompose), and are recyclable. A primary benefit in using thermoplastic matrices in PMCs over thermosets is, that they can be remoulded through heating.

References

[1] Ji H W, Koo D D and Kang J-H 2020 Short- and long-term structural characterization of cured-in-place pipe liner with reinforced glass fiber material *Int. J. Environ. Res. Public Health* **17** 2073

[2] Zhang C, Wang X, Liu A, Pan C, Ding H and Ye W 2021 Reduced graphene oxide/titanium dioxide hybrid nanofiller-reinforced electrospun silk fibroin scaffolds for tissue engineering *Mater. Lett.* **291** 129563

[3] Ma X, Wu G, Dai F, Li D, Li H, Zhang L and Deng H 2021 Chitosan/polydopamine layer by layer self-assembled silk fibroin nanofibers for biomedical applications *Carbohydr. Polym.* **251** 117058

[4] Yuan J, Xing Z C and Park S W *et al* 2009 Fabrication of PHBV/keratin composite nanofibrous mats for biomedical applications *Macromol. Res.* **17** 850–5

[5] Kim H M, Chae W P, Chang K W and Chun S 2010 Composite nanofiber mats consisting of hydroxyapatite and titania for biomedical applications *J. Biomed. Mater. Res.* **94** 380–7

[6] Wanna D, Alam C, Toivola D M and Alam P 2013 Bacterial cellulose–kaolin nanocomposites for application as biomedical wound healing materials *Adv. Nat. Sci. Nanosci. Nanotechnol.* **4** 045002

[7] Veliz D S, Alam C, Toivola D M, Toivakka M and Alam P 2014 On the non-linear attachment characteristics of blood to bacterial cellulose/kaolin biomaterials *Colloids Surf.* **B116** 176–82

[8] Li C, Huang Y and Li R *et al* 2021 Fabrication and properties of carboxymethyl chitosan/polyethylene oxide composite nonwoven mats by centrifugal spinning *Carbohydr. Polym.* **251** 117037

[9] Chawla K K 2019 Carbon fiber/carbon matrix composites *Composite Materials* (Cham: Springer)

[10] Naslain R R and Pomeroy M R 2016 Ceramic matrix composites: matrices and processing *Reference Module in Materials Science and Materials Engineering*

[11] Donald I W and McMillan P W 1976 Ceramic matrix composites *J. Mater. Sci.* **11** 949–72

[12] Chawla K K 1993 *Ceramic Matrix Composites* (Berlin: Springer)

[13] Pank D R and Jackson J J 1993 Metal-matrix composite processing technologies for aircraft engine applications *J. Mater. Eng. Perform.* **2** 341–6

[14] Sapuan S M 2017 *Composite Materials–Concurrent Engineering Approach* (Oxford: Butterworth-Heinemann)

Parvez Alam

Chapter 14

'N'

14.1 Natural fibre composite (NFC) materials

Natural fibre composite (NFC) materials are composites with a reinforcing phase made up of predominantly, natural fibres originating from renewable wood or plant sources. They are already widely used in cars for their acoustic damping properties, as car interior material, in brake pads, in construction, in boats, and in cycle frames (often as a hybrid with carbon fibre). Natural fibres can be divided into the five broad groups listed below:

- **Woods:** softwoods and hardwoods.
- **Stem/Bast fibres:** flax, jute, hemp, kenaf, rattan, ramie, okra, kudzu.
- **Leaf fibres:** sisal, abaca, pineapple, banana, fique, henequen, palm, cantala, phormium.
- **Seed/Fruit fibres:** cotton, coconut.
- **Grass fibres:** bamboo, rice, miscanthus giganteus.

Natural fibres composites may be incorporated into thermoplastic or thermoset matrices. These may be biodegradable or non-biodegradable and could be from renewable sources (biological sources) or synthetic (crude oil based). Many biodegradable polymers may undergo photo biodegradation (polymer degradation onset by the presence of UV light) assisted by UV sensitisers. The polymer is broken down into lower molecular weight segments and then converted to CO_2 and H_2O using specialised bacteria (a process known as mineralisation). Biodegradable plastics may also be composted using specialised microbials (bacteria, fungi, algae). A few examples of natural polymer matrix materials include:

- **Biodegradable:** gluten, starches, corn based thermoset polyester, soy protein resin.
- **Non-biodegradable:** cashew nut shell liquid, wood based epoxides, furan resin.

doi:10.1088/978-0-7503-3527-0ch14

NFCs are often considered potential natural replacements for synthetically manufactured fibre composites. A major hurdle with this is that natural fibres are still comparably low in terms of their mechanical performance when compared against synthetic fibres, as noted in table 14.1, which summarises a few select physical and mechanical properties of synthetic and natural fibres. To improve mechanical performance, natural fibres can be subjected to different types of chemical treatments. The more common ones are listed below, however, recent efforts to skeletise fibres using templated ceramics [1] and bio-glasses [2] show that considerable improvements in mechanical properties are achievable, though these research efforts are still in their early stages.

- **Acetylation:** the process of combining cellulose with acetic acid and thus introducing an acetyl radical into an organic molecule.
- **Bleaching:** a method for improving the whiteness of textile fibres.
- **Coating with polymeric solutions:** such as phenol-formaldehyde or resorcinol formaldehyde coatings on jute fibres.
- **Grafting:** integration of monomers or oligomers (short chain polymers) to the fibre surface by chemical reaction. Jute for example can be graft co-polymerised with vinyl monomers (e.g. methyl methacrylate, ethyl acrylate, styrene, vinyl acetate).
- **Mercerisation:** (from John Mercer, 1844) use of concentrated caustic soda (alkali—sometimes referred to as alkalisation). The fibres are swollen and the strength/dye affinity of the fibres are increased.
- **Scouring:** or, solvent treatment. The treatment of natural fibres in solvents to remove natural fats, waxes, proteins, dirt, oil and other impurities.

Table 14.1. General properties of natural fibres compared against a few common synthetic fibres, here ρ is density (kg m^{-3}), E is the tensile modulus (GPa), σ is the tensile strength (MPa), and ε is the elongation (%) [3].

Category	Fibre	ρ	E	σ	$\frac{E}{\rho}$	$\frac{\sigma}{\rho}$	ε
Synthetic fibres	E-glass	2550	71	3400	0.028	1.333	3.4
	S-glass	2500	85	4580	0.034	1.832	4.6
	Carbon	1400	235	4000	0.168	2.857	2.0
	Aramid	1440	124	2760	0.086	1.917	2.5
Bast fibres	Jute	1520	60	860	0.039	0.566	2.0
	Hemp	1520	70	920	0.046	0.605	1.7
	Flax	1520	100	840	0.066	0.553	1.8
	Ramie	1500	25	560	0.016	0.373	2.5
	Kenaf	1300	50	700	0.038	0.538	2.1
Leaf fibres	Sisal	1450	22	640	0.015	0.441	7.0
	Pineapple	1440	82	1627	0.057	1.130	1.6
Seed fibres	Coir (coconut)	1150	6	175	0.005	0.152	40.0
	Cotton	1520	27	800	0.018	0.526	12.0
Grass fibres	Bamboo	1400	50	740	0.036	0.529	2.0
	Giant miscanthus	240	3.7	28	0.015	0.117	

14.2 Nanocomposites

Nanocomposites refer to composites containing one or more components with dimensions below 100 nm. Nanocomposites may also refer to material structures that have nanoscale repeat units between the different material phases. Common nano-reinforcement types include; particles, sheets/stacks, core–shell, yoke shell, whiskers, fibres and nano-inclusions, and these are commonly used to optimise the optical, electrical, thermal and mechanical properties of composite materials. Figure 14.1 illustrates a few examples of different categories of nanomaterial used in nanocomposite membranes for gas and liquid separation. These are characterised by their geometries and dimensions as 0D, 1D, 2D and 3D. 0D nanomaterials are effectively spherical with all dimensions below 100 nm, 1D nanomaterials have one single dimension above 100 nm, 2D nanomaterials have two Cartesian dimensions above 100 nm and 3D nanomaterials include nanofeatures within a 3D structure with all continuum dimensions above 100 nm.

Nanocomposites behave differently to normal composites because they have a very high surface area to volume ratio (typically one order of magnitude higher). As a consequence, small quantities of reinforcement can have highly noticeable effects

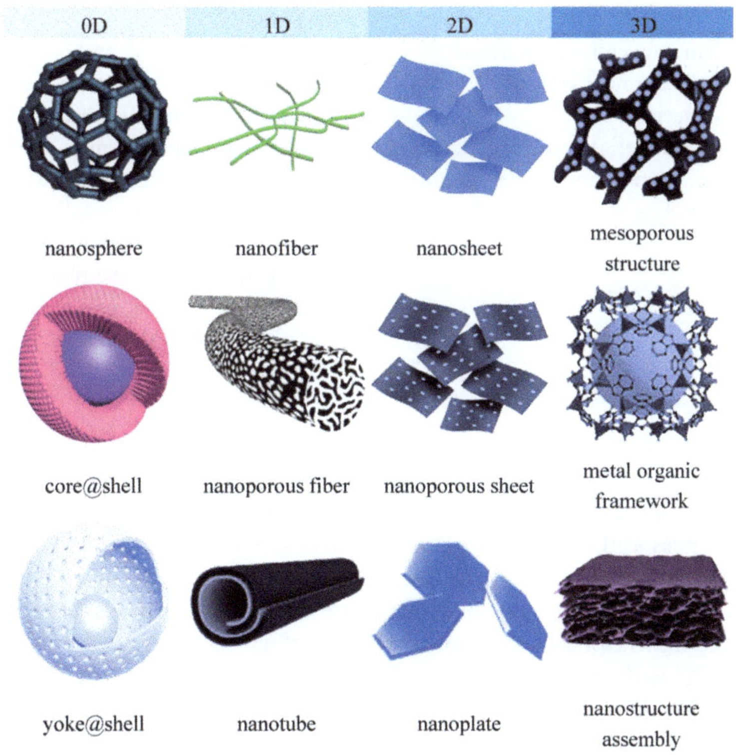

Figure 14.1. Examples of 0D, 1D, 2D and 3D nanomaterials used in gas and liquid separation as part of a nanocomposite membrane. Reprinted from [4] with permission from MDPI (CC BY 4.0).

Table 14.2. A few examples of benefits and potential drawbacks of using nanoscale materials in nanocomposites.

Benefits	Potential drawbacks
Excellent strength, stiffness and durability	Exposed nanosurfaces can be highly reactive.
Often lightweight	Ecotoxicological and toxicological effects may arise on exposure to nanoscale materials. Exposure for humans may be dermal (skin products containing nanoparticles), through inhalation, by ingestion (food products contained in nanocomposite packaging).
Less inclusion/reinforcement material usage	Free nanoparticles may enter into the environment through composite degradation. Direct passage into the environment may occur via waste water, spills, transportation, during production and disposal. When in the environment, nanoparticles may enter the food chain by plants, animals and humans.
Less total material volume in manufacture and as waste/degrading/recycling	The small size of nanoparticles, allows them to, in some instances, bind to proteins thus changing their form and the way in which they may act.
Less energy than conventional material equivalents on combustion due to lower volume of material	Nanoparticles based on carbon are lipophilic and can thus accumulate in the cells of biological organisms (bioaccumulation).
Many green nanomaterials can be used (e.g. nanocellulose, starch nanoneedles, chitin nanoneedles etc)	If inhaled, nanoparticles may give rise to inflammations of the respiratory tract and may even damage it.
	Nanoparticles are also small enough to be easily transported through the blood stream to different organs and may accumulate/react/cause mechanical damage of sorts.

on the global properties of these composites. There can be significant cost reductions through the use of nanoscale reinforcements. This depends of course on the type of reinforcement. Nano-clay platelets will be relatively cheap and can provide superior strength, stiffness and fracture toughness. Carbon nanotubes however, often used to improve electrical and thermal conductivity, are considerably more expensive.

In a two-phase nanocomposite where $_n$ represents the nanoparticle phase and $_m$ represents the matrix phase, the energy of the surface, U_s, can be defined by equation (14.1), where U is the internal energy of the system, U_n and $(U_n + U_m)$ represents the total uniform energy available in both nanoparticle and matrix phases. It should be noted that surface pinning creates an interphase between the nanoparticle and the matrix, and as such, U_s essentially represents the energy of the interphase in as much as it does the energy of the surface when the surface is assumed to include the interphase.

$$U_s = U - (U_n + U_m) \tag{14.1}$$

When force is applied to a nanocomposite, work is done at the surface, W_s and this is represented by equation (14.2), where γ is the surface stress and A is the area of the surface. A volumetric change results from surface deformation such that $\mathrm{d}V_n = -\mathrm{d}V_m$ where V represents volume. The deformation at the surface can additionally be expressed in the form of a Lagrangian strain at the surface such that $\varepsilon_s = \frac{\mathrm{d}A}{A}$ which relates then to the surface stress γ as shown in equation (14.3), where E_s is the free energy of the surface [5].

$$W_s = \gamma \mathrm{d}A \tag{14.2}$$

$$\gamma = E_s + \frac{\partial E_s}{\partial \varepsilon_s} \tag{14.3}$$

While there are advantages to using nanoscale inclusions in composites, there are also several health and environmental drawbacks. A few benefits and disadvantages are listed in table 14.2.

References

[1] Alam P, Fagerlund P, Hagerstrand P, Toyryla J, Amini S, Tadayon M, Miserez A, Kumar V, Phalevan M and Toivakka M 2015 L-lysine templated $CaCO_3$ precipitated to flax develops flowery crystal structures that improve the mechanical properties of natural fibre reinforced composites *Composites* **A 75** 84–8

[2] Zusron M *et al* 2016 Glass coating natural fibres by diatomisation: a bright future for biofouling technology *Mater. Today Commun.* **7** 81–8

[3] Alam P 2010 *Understanding Composites: Engineering and the Environment* unpublished class notes from the author's (Parvez Alam) lecture series on composites held at Abo Akademi University, Finland

[4] Goh P S, Wong K C and Ismail A F 2020 Nanocomposite membranes for liquid and gas separations from the perspective of nanostructure dimensions *Membranes* **10** 297

[5] Alam P 2015 Mechanical properties of bio-nanostructured materials *Handbook of Mechanical Nanostructuring* ed M Aliofkhazraei 1 (Weinheim: Wiley) pp 211–35

IOP Publishing

Composites Engineering: An A–Z Guide

Parvez Alam

Chapter 15

'O'

15.1 Orowan strengthening (MMCs)

Reinforcing particles within a metal matrix can strengthen the matrix by blocking dislocation movement. A well-known mechanism describing this is the Orowan mechanism. This mechanism assumes that a dislocation moves between two non-shearing particles orthogonally to an intersecting line between the particles such that the apex of the bow aligns with the centre of the intersecting line. The Orowan mechanism consists of three main stages. In the first stage, the moving dislocation begins to bow between the two particles. In stage two, local yielding occurs as the dislocation bow approaches a semi-circle in its shape. In the final stage, yielding has occurred, and the dislocation leaves Orowan (dislocation) loops around the particles as the dislocation is unable to overcome the phase boundary between the metal matrix and the particles. The mechanism is illustrated in figure 15.1.

The shear stress can be predicted by equation (15.1), where G is the shear modulus, $\mathbf{b}$ is the Burgers vector of the dislocation, and γ is the interparticle spacing, represented by equation (15.2). In this equation, L is the distance between the particle centroids and r is the radius of the particles. It can be noted that τ increases as $\sqrt{r}$ and resistance (hardening of the MMC) is a function of particle size.

$$\tau = \frac{G\mathbf{b}}{\gamma} \tag{15.1}$$

$$\gamma = L - 2r \tag{15.2}$$

Equation (15.1) is the basic form describing the Orowan mechanism and the resultant shear stress. It does not take into account any changes to the dislocation itself as a result of the bowing. As such, equation (15.3) can be used for a more accurate calculation of shear stress, where θ is the angle between the Burgers vector and the dislocation line, A is the matrix-particle surface area and is expressed as a function of θ, and L_{eff} is the effective particle separation.

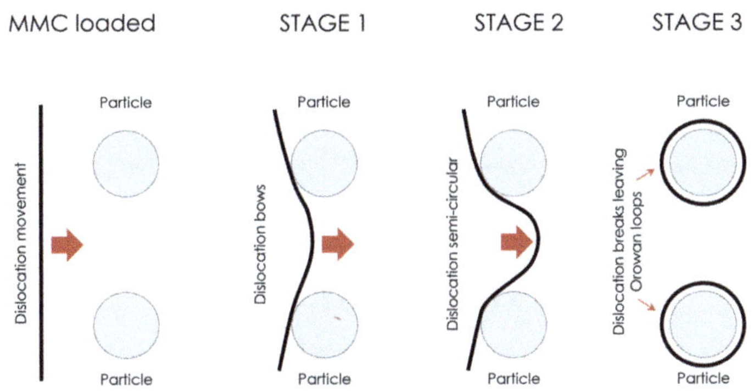

Figure 15.1. Illustrative scheme of the Orowan mechanism.

$$\tau = A(\theta)\frac{G\mathbf{b}}{2\pi L}\ln\left(\frac{L_{\text{eff}}}{r}\right) \tag{15.3}$$

Orowan yielding $\Delta\sigma_{y,\text{Orowan}}$ can be generally described by equation (15.4), where C^* and C are empirical constants, d is the particle diameter, and X is a numerical factor associated with the particle size typically between 1 (particle diameter) and 2 (particle radius) [1].

$$\Delta\sigma_{y,\text{Orowan}} = C^*\frac{G\mathbf{b}}{L}\ln\frac{d}{X\mathbf{b}} = C\frac{G}{\frac{L}{d}}\left(\frac{X\mathbf{b}}{d}\right)\ln\frac{d}{X\mathbf{b}} \tag{15.4}$$

Reference

[1] Ferguson J B, Lopez H and Kongshaug D *et al* 2012 Revised Orowan strengthening: effective interparticle spacing and strain field considerations *Metall. Mater. Trans.* **A 43** 2110–5

IOP Publishing

Composites Engineering: An A–Z Guide

Parvez Alam

Chapter 16

'P'

16.1 Physical vapour deposition (PVD) methods in MMCs

Physical vapour deposition (PVD) is a method for manufacturing metal matrix composites (MMCs). The process is conducted in a chamber with gas inlets and outlets, containing a target cathode and an anodic work piece. During PVD, fibres are passed through the chamber containing vapourised (evaporated) metal matrix.

A thin or thick film of the metal matrix will be deposited through condensation of the vapour upon the fibre surface. The metal matrix vapour is generated using high-energy electrons that are directed at a matrix feedstock in its solid form. The fibres now coated by a layer of metal matrix are then collected and formed into the final MMC form by pressing under elevated temperatures. Alternatively, a sputtering technique can be used, which involves bombarding the metal feedstock with ions from a processing gas such as argon, after which the coated fibres can be hot pressed or hot isostatically pressed to create the final MMC. Figure 16.1 shows a scheme of the PVD method based on the sputtering method.

Deposition rates can be up to 10 μm min^{-1} using the sputtering method, or up to 3–600 mm min^{-1} using the high-energy electron beam (though only about one tenth of this actually coats the fibre) [1]. One of the recognised benefits of PVD is that since the metal feedstock is vapourised, alloys and combinations of metals can be vapourised to enable the physical deposition of the condensed combination of vapours to the surfaces of the fibres.

16.2 Piezoelectric composites

Piezoelectric materials (PZTs) are able to convert mechanical to electrical energy. Under a mechanical deformation, a piezoelectric material will generate an internal charge. This happens due to polarisation, P, in a piezoelectric crystal, which is linearly proportional to the stress, σ, that produced it such that $P = d\sigma$, where d is the piezoelectric coefficient. The piezoelectric effect is due to the noncentrosymmetrical

doi:10.1088/978-0-7503-3527-0ch16

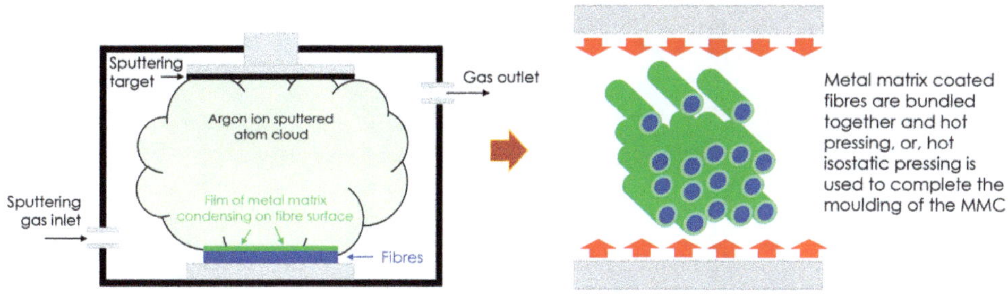

Figure 16.1. Scheme of the physical vapour deposition (PVD) process for the manufacture of metal matrix composites (MMCs).

crystal structure of piezoelectric crystals that allows for stress-activated atomic shifts, resulting in turn, in polarisation.

16.2.1 Piezoelectric charge constant, *d*

In piezoelectric composites, polarisation is generated when the composite is subjected to mechanical stress. This is reversible, such that mechanical deformation (or strain) can be induced when an electric field is applied to the composite. This phenomenon, both normal and reversible, is termed the piezoelectric charge constant and is represented by d_{ij}, where $_i$ is the direction of polarisation when the electric field is 0, or the direction of an applied electric field, and $_j$ is the direction of applied stress, or the direction of induced strain. There are three main constants, d_{33}, d_{13} and d_{15}. Here, the numerical values 1 and 3 refer to the first and third directions of local orthogonal symmetry in the material, and 5 refers to shear about axis 2.

16.2.2 Piezoelectric voltage constant, *g*

In piezoelectric composites, an electrical field is generated when the composite is subjected to mechanical stress. This is reversible, such that mechanical deformation (or strain) can be induced when a unit of electric displacement is applied to the composite. This phenomenon, both normal and reversible, is termed the piezoelectric voltage constant and is represented by g_{kl}, where $_k$ is the direction of the electric field generated, or the direction of an applied electric displacement, and $_l$ is the direction of applied electric displacement, or the direction of induced strain. There are three main constants, g_{33}, g_{13} and g_{15}. Here, the numerical values 1 and 3 refer to the first and third directions of local orthogonal symmetry in the material, and 5 refers to shear about axis 2.

16.2.3 Dielectric constant, *ε*

The dielectric constant, ε, also referred to as the permittivity constant, is the dielectric displacement under an electric field. There are two constants of importance, each of which appears as ε_{mn}^{X}. Here, X may refer to either constant stress, ^{T}, or

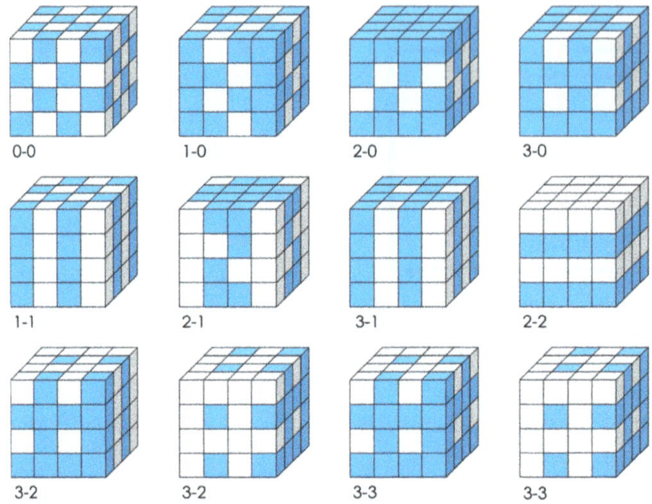

Figure 16.2. 10 variants of connectivity in two-phase PZT:polymer composites (note that 3-2 and 3-3 exist in two configurations and thus there are 10 variants rather than 12).

constant strain, ^{S}, while the subscript $_m$ indicates the direction of the dielectric displacement and $_n$ indicates the direction of the electric field. The two constants commonly referred to are ε_{11}^T and ε_{33}^S.

16.2.4 Connectivity of active-passive phases in piezoelectric composites

Piezoelectric composites are combinations of piezoelectric ceramics with polymerics (PZT:polymer composites). The piezoelectric ceramic is termed an active phase (m) while the polymer is termed a passive phase (n). The relationship between active and passive phases ($m - n$) is called connectivity. Connectivity is a simple way of describing the connections between active and passive phases through the composite, as importantly, this will influence the way in which a PZT:polymer composite will directionalise the output of electrical energy. A single phase connected in three orthogonal axes (x, y and z) is termed '3'. A single phase connected in only one axis is termed '1'. If connectivity ($m - n$) is then, 0-0, it means that neither the active nor passive phases are connected, while 1-3 means that the active phase is connected in one axis, while the passive phase is connected in 3 axes. In a two-phase (diphasic) PZT:polymer composite, there are 10 possible variations [2], 0-0, 1-0, 2-0, 3-0, 1-1, 2-1, 3-1, 2-2, 3-2, 3-2, 3-3, 3-3, note that the repetition of 3-2 and 3-3 refers to two different configurations in each hence 12 variations are in fact 10, figure 16.2.

16.2.5 Effects of combining PZT with polymers on piezoelectric constants

While the above constants are true for neat PZT materials, when combined in the form of a composite the overall effect of these constants in the active PZT phase is altered by the presence of a passive polymer phase. Uchino [2] details the properties of PZT composites against neat single phase PZT material (PZT501A—lead titanium zirconium oxide). From this we can deduce that ε of PZT:epoxy (connectivity 3-1), PZT:silicone rubber (connectivity 3-3), PZT:rubber (connectivity 3-0) is reduced to 20%, 2% and 3.65% from the neat PZT, respectively. Further, d_{33} of PZT:epoxy (connectivity 3-1), PZT:silicone rubber (connectivity 3-3), PZT:rubber (connectivity 3-0) is reduced to 75%, 27.5% and 13% from the neat PZT, respectively. Finally, g_{33} of PZT:epoxy (connectivity 3-1), PZT:silicone rubber (connectivity 3-3), PZT:rubber (connectivity 3-0) is increased to 375%, 1400% and 700% from the neat PZT, respectively.

While there are other constants used in PZT composites engineering they align with other 'mechanical' engineering constants, e.g. stiffness, compliance etc, and will therefore not be focussed on here.

16.3 Polymer impregnation and pyrolysis (PIP) in CMCs

Polymer impregnation and pyrolysis (PIP), also often referred to as polymer infiltration and pyrolysis (PIP), is a method by which means ceramic matrix can be formed in ceramic matrix composites (CMCs). The PIP process begins with a fibre preform which is impregnated by an organic precursor material that can be pyrolysed to form a ceramic matrix. This organic matrix is not a typical thermoset polymer used in polymer matrix composites (PMCs), such as an epoxy resin, but is specially chosen to enable the formation of a specific ceramic matrix. For example phenolic thermoset precursors are used to generate a carbonaceous matrix, polycarbosilane precursors are used to form silicon carbide matrices, and polycarbosilazane precursors are used to form Si–C–N matrices [3]. Once the preform is impregnated, the organic matrix is solidified by curing using a relevant curing method, e.g. chemically catalysed or by heat. The polymer component of the PMC is then pyrolysed at ca. 700 °C–1000 °C and the ceramic matrix forms within the preform to create the CMC. Once pyrolysed, stabilisation of the ceramic matrix can then be undertaken at higher processing temperatures. Since the ceramic yield is not always high, the process may need to be repeated several times to increase the ceramic matrix fraction of the composite. Parameters of importance in this manufacturing process include the pyrolysis temperature, the heating rate, the dwell time and the temperature at which the microstructure is stabilised. The PIP process has several advantages including:

- the process is relatively simple and does not require much specialised machinery,
- ceramic matrices can be formed at relatively low temperatures thus sparing damage to fibres,
- there is versatility in the types of reinforcements that can be used,
- the method leads itself to a broad range of ceramic matrices derived from liquid organic precursors,
- the density and microstructure of the ceramic microstructure can be controlled,
- net shape parts can be manufactured.

There are nonetheless drawbacks to the PIP process and these include:

- manufacture times can be lengthy due to repeat precursor impregnation/infiltration,
- organic precursor choices should also include good wetting properties to both reinforcement and the pyrolysed ceramic matrix,
- the process can lead to porosity when ceramic yield is sufficiently low, or follow-on precursor impregnation is unable to permeate and wet the microstructure properly,
- overall production costs are high compared to other infiltration processes (e.g. melt infiltration) [4].

16.4 Porous composites

Porous composites are different from porous materials. While *porous materials* are essentially single material phases with porosity, *porous composites* are two-phase materials, with interfaces between the phases and porosity within the composite structure. Porosity is ordinarily considered to the detriment of composite materials and are categorised as defects. In porous composites however, porosity is desired and is a purposeful element of the composites design process. There is a broad range of application of porous composites including for example—coatings, storage devices, sieves, sensors, as breathable and cell-interlocking biomedical materials, lightweight structural applications, and as insulators. The desired characteristics of porous composites are thence mass transfer, fluid retention, opacity, heat transfer, architecture, and density.

16.4.1 Types of pores and their geometrical characteristics

There are two main categories of pore: open pore and closed pore. Depending on the specific field within composites engineering these are sometimes also referred to as open cell and closed cell (e.g. in foams/foam cores and honeycombs and metamaterials). Pores may also be connected and closed or connected and open, or disconnected and closed. If pores are disconnected, the simplifying assumption is often made that the pore can be approximately defined by a geometrical primitive, such as a spheroid, an ellipsoid, a cube or a cuboid. If a pore is connected (whether open or closed) the pore space is often subdivided into pore-parts to enable easier identification of understanding of the pore space. Connected pores are commonly subdivided into the pore (or pore stomach), the pore throat, and the pore neck. Figure 16.3 illustrates the above concepts.

16.4.2 Particle packing

Porosity can occur passively in certain composites, e.g. paper coatings (comprising particulates, binder and air). When this happens the particle sizes, shapes and structures influence the developing pore space microstructures. An understanding of factors related to particle packing can be useful in such instances. Idealised monodisperse spherical particle packings assuming no binder material include cubic packings, hexagonal close packings, orthorhombic packings and rhombohedral

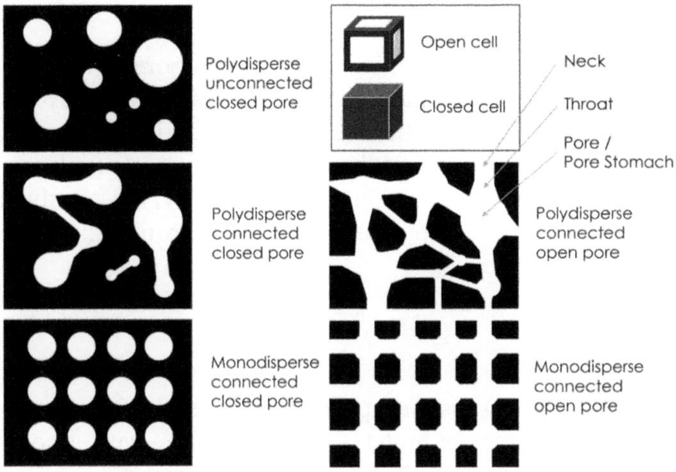

Figure 16.3. Open and closed pore spaces (cells) for both monodisperse and polydisperse pores. Connected and unconnected pores are shown as is the dicretisation of a pore into a neck, throat and pore/pore stomach.

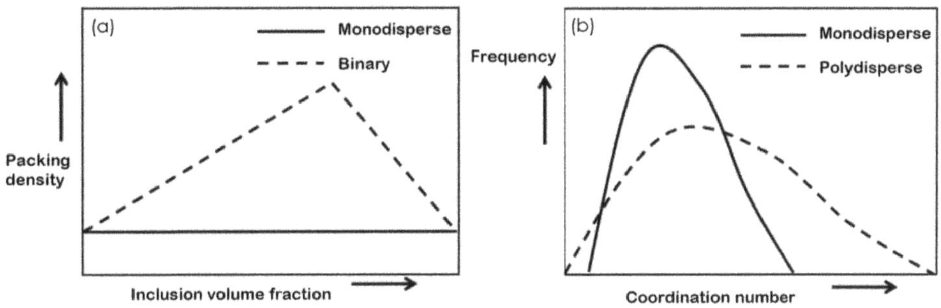

Figure 16.4. Idealised graphical schemes showing (a) packing density functions of monodisperse and binary packings, and (b) comparisons of the coordination number frequencies for monodisperse and polydisperse particle systems. Reprinted from [5] InTech Open (CC BY 4.0).

packings. Binder wets and adheres adjacent particulates and as a result, idealised packings such as these are seldom seen in such composites. Polydisperse spherical particles have higher packing densities than monodisperse packings as smaller particles can fit interstitially between larger particles, filling up therefore, the negative space between larger particulates. This can be more clearly understood from figure 16.4(a), where binary particle mixtures are noted to be able to reach an apex in packing density, while monodisperse particles retain a constant density at all particle fractions. Polydisperse systems also tend to have greater flexibility in terms of the coordination numbers (numbers of nearest neighbour particles) since smaller interstitial particles can fit between larger particles, figure 16.4(b). This can affect both the pore microstructure and the solid state continuity of the composite.

16.4.3 Permeability

Permeability is possible in composites containing open pores and porous composites are often designed specifically for mass transfer applications. Darcy permeability is a good model that works well at low flow rates, under laminar flow, and assuming an incompressible fluid, equation (16.1), where Q is the flow rate (volume flux) through the composite, A is the area, K is the Darcy permeability, μ is the dynamic viscosity of the penetrating fluid, ΔP is the pressure difference through the composite and Δx is the thickness of the composite.

$$Q = -\frac{KA}{\mu}\frac{\Delta P}{\Delta x} \tag{16.1}$$

The use of equation (16.1) assumes porosity is at least relatively homogeneously dispersed through the composite material, and that flow with a constant dynamic viscosity is, ultimately, a function of the pressure difference.

16.4.4 Elastic modulus predictions

A number of models can be used to predict the elastic modulus of porous composites. Equation (16.2) simply assumes that the elastic modulus, E, is inversely proportional to the fraction of porosity, ϕ. Here, E_c is the composite elastic modulus assuming no porosity.

$$E = E_c(1 - \phi) \tag{16.2}$$

Equation (16.3) is a semi-empirical model that is similar to equation (16.2), except that there is an porosity is modified by the fitting parameters a and b.

$$E = E_c(1 - a\phi^b) \tag{16.3}$$

McAdam's model [6], equation (16.5), replaces E_c in equation (16.2) with the transverse rule of mixtures model (Reuss model) for particulate composites containing pores. Here, V_r and V_m are the volume fractions of reinforcement and matrix, respectively, while E_r and E_m are the elastic moduli of the reinforcement and matrix, respectively.

$$E = \left(\frac{V_r}{E_r} + \frac{V_m}{E_m}\right)^{-1}(1 - \phi) \tag{16.4}$$

Bert's model [7] modifies the Halpin–Tsai model for particulate composites containing pores as shown in equation (16.8), where ξ is a shape specific parameter for the reinforcement, and η is calculated using equation (16.6). For fibres or particles with either approximately circular or square cross sections (i.e. where there is an aspect ratio close to unity), $\xi = 2$. ξ can be determined according to equation (16.7), where a is the dimension of the reinforcement in the direction of loading, while b is the dimension of the reinforcement transverse to the direction of loading.

$$E = \frac{E_m(1 + \xi\eta V_r)}{(1 - \eta V_r)}\left(1 - \frac{\phi}{(1 - V_r)}\right)^{K_0(1 - V_r)} \tag{16.5}$$

$$\eta = \frac{\dfrac{E_r}{E_m} - 1}{\dfrac{E_r}{E_m} + \xi} \tag{16.6}$$

$$\xi = 2\left(\frac{a}{b}\right) \tag{16.7}$$

Alam's model [8], equation (16.8) has been shown to be very accurate for particle-polymer composites containing porosity and near unity aspect ratios. This model use the rule of mixtures model as a baseline and proposes that it be modified to account for the effective binder fraction in contact with particulates $V_{m,\text{eff}}$, a stress transfer aspect ratio s_t, a reinforcement aspect ratio $\frac{L_r}{d_r}$ where L_r and d_r are the average length and diameter of the reinforcement, respectively, an anfractuosity parameter A, and an effective pore width ratio $\frac{w_{p,\max}}{w}$ where $w_{p,\max}$ is the maximum pore width measured over a composite width under scrutiny, w. If there are no pores, $A = 1$ and $w_{p,\max} = 0$.

$$E = \left(E_r V_r\left[\frac{L_r}{d_r}\cdot\frac{1}{s_t}\right] + E_m V_{m,\text{eff}}\right)\cdot\frac{1}{A}\cdot\left(a - \frac{w_{p,\max}}{w}\right) \tag{16.8}$$

16.5 Post-curing

Post-curing is a method used to improve the mechanical properties of a cured composite. There are a variety of post-curing methods for a variety of types of composites, each of which will be detailed individually.

16.5.1 Heat post-curing of polymer matrix composites

Heat is the most common method for post-curing polymer matrix composites (PMCs) comprising thermoset resin matrices. The method is sometimes used to speed up the curing process of room-temperature (RT) cured PMCs manufactured using out-of-autoclave (OoA) processes. An OoA processed composite may take over a week to achieve full cure at RT and thus to exacerbate the curing process, post-curing for RT cured OoA composites will be typically instigated after the first 12 h of RT cure. Post-curing such parts allows the composite part to reach a maximum cross-link density at a much faster rate (ca. 24 h). Post-curing should be conducted for RT cured OoA composites prior to machining. A part that is RT cured under its RT complete-cure time (> a week) may experience additional curing, leading to warping or shrinkage and a loss in dimensional stability. It should be noted that if an RT cured OoA composite has already completed its full cure cycle, there will be neither benefit nor harm in post-curing the composite.

Post-curing is also conducted on composites that have already been heat-cured, but which have not reached their full cross-link density. This can quickly and easily be ascertained by running a sample of the composite through a differential scanning calorimeter (DSC), which will indicate the curing/non-curing as heat is elevated in the machine. Post-curing heat-cured composites has been shown in many cases to significantly improve mechanical performance.

There are various heating methods that are used to cure PMCs. Radiant heaters produce long wave infrared radiation and radiant heating is a beneficial method for post-curing as it does not require any specific heating enclosure. It is a useful technique on large composite parts and components which would otherwise be expensive to make oven enclosures for. A disadvantage in using radiant heaters is that temperature control is imprecise and temperatures are moreover difficult to maintain. As such, there is always a danger with this method that parts of the composite part will not be fully cured and this will have a knock-on effect on mechanical performance and part reliability. The more reliable heating methods involve the use of a fully insulated enclosed oven with either electric, oil or vented gas heating. Post-cured parts should when possible maintain some form of pressure to avoid possible bubbling or warping during the post-cure.

16.5.2 Light post-curing of polymer matrix composites

Light curing of PMCs is often used in the dental practice, where curing takes place at the site of the tooth and heat treatments are potentially hazardous to the patient's health. Special light activated resins are used to enable this type of post-curing to take place. To date UV light, halogen bulbs, laser, plasma, and light emitting diodes (LEDs) have been used to post-cure dental composites. It should be noted that of these, UV post-curing has been discontinued for dental composites as the waves are too long to penetrate sufficiently deep into the composite (shorter wavelengths of light are needed) and this has been replaced by the halogen bulb. These dental post-cure lights fall into the category of 'blue light'. While the plasma light is relatively fast to use, it is also expensive to maintain. LEDs are the most popular blue light post-curing method for dental composites. While UV lights are no longer used to post-cure dental composites, they are used to post-cure additive manufactured resin prints, which is an area that is gaining interest from composites specialists for the complex and controllable cellular solid parts that can be made.

16.5.3 Microwave post-curing of polymer matrix composites

Accelerated post-curing of thermoset composites (mainly epoxy, polyester and vinyl ester matrix composites) can be conducted using microwave radiation. While the process is understood to reduce post-curing time and styrene emissions, the process is challenging and parameters of importance include penetration depth, dielectric properties and the microwave frequencies used for post-cure.

16.6 Preform

Preforms are dry fibre assemblages that are pre-shaped into the desired form of a composite product. They can therefore be viewed as the reinforcing shaped endo-skeleton intended for the final consolidated composite part. When using preforms, it is important to factor in changes not only to the fibre orientation due to preforming curvatures, but also to factor into the preform design, potential increases in waviness, especially at mould corners. Preforming is conducted using either hand-lay techniques, or is automated using robotic tools, often with interchangeable end effectors to enable varied functionality. Preforms can be broken down by their final form and this section therefore aims to introduce the reader to common 1D, 2D and 3D preforms [9].

16.6.1 1D preforms

One dimensional preforms are essentially yarns, or linear assemblies of individual fibres. There are two main types of yarn used in both single type and co-mingled type, these being spun yarns and continuous filaments. Table 16.1 provides different examples of yarns and co-mingled yarns that may be used in 1D preforming.

16.6.2 2D preforms

There are several types of 2D preforms including uni-directional layered, woven textiles, braided fabrics, knitted fabrics and random fibre mats. Table 16.2 provides a few examples from each group.

16.6.3 3D preforms

There are several types of 3D preforms that enable the manufacture of complex 3D shaped composites. Table 16.3 provides a few examples.

16.7 Pultrusion

Pultrusion in FRP manufacture is a process that begins with the raw materials of fibre and resin and ends with shaped composite objects that have a continuous z-direction dimension and can be cut to size. Examples of pultruded composite

Table 16.1. Generalised types of 1D preforms and accompanying examples.

Type	Examples
Single	Z-twist spun yarn S-twist spun yarn Linear continuous filament yarn Twisted continuous filament yarn
Co-mingled	Intermingled yarn Co-wrapped yarn Core-spun yarn Stretch-breaking yarn

Table 16.2. Generalised types of 2D preforms and accompanying examples.

Type	Examples
Uni-directional layered	Single layer tapes Multi-layered tape (uni or multi-directional) Stitched uni-directional tows Knitted uni-directional tapes
Woven textiles	Bi-axial plain weave Bi-axial basket weave Bi-axial twill weave Bi-axial satin (sateen) weave Tri-axial weave Quadraxial weave Unbalanced leno weave Unbalanced quasi uni-directional weave
Braided fabrics	Bias-braids (flat, tubular, square) Tri-axial braids (flat, tubular, square)
Knitted fabrics	Plain knit Plain knit with weft-inlay Double pique Interlocked Variable half-milano Interlock with weft-inlay Uni-axial inlay Bi-axial inlay Orthogonal bi-axial inlay
Random	Chopped strand mats Continuous filament random mats Surface veils Shaped chopped strand preforms
Non-crimp fabrics	Bi-axial Tri-axial Quadraxial

shapes are shown in figure 16.5. Pultruded parts are commonly used in small-scale civil structures such as bridges and buildings.

The pultrusion process is almost entirely automated, producing continuous lengths of pultruded composite material with constant cross/sections, and often with high fibre volume fractions. Almost any fibre type can be used in the process, and several resin systems are commonplace including epoxies, polyesters, vinyl esters and phenolics. A schematic of the process is shown in figure 16.6. Initially, reinforcement, commonly fibre or wovens, is pulled in tension by rollers and fed directly into a resin containment where the reinforcement is soaked by the resin.

Table 16.3. Generalised types of 3D preforms and accompanying examples.

Type	Examples
Solid textiles	Traditional 3D wovens Orthogonal non-wovens 3D braids 3D knits Stitched fabrics
Woven sandwich structures	Velvet weave
Knitted sandwich structures	Closed skin 3D knitted sandwich Open skin 3D knitted sandwich

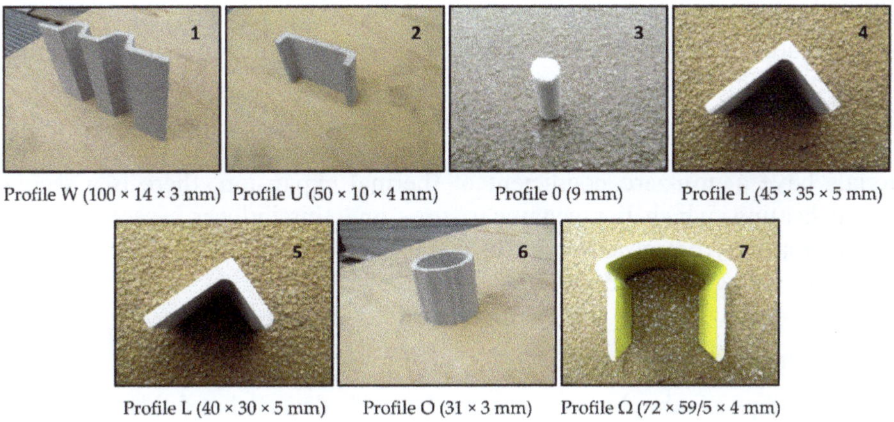

Figure 16.5. A few examples of pultruded glass fibre reinforced plastic sections. Reprinted from [10] with permission from MDPI (CC BY 4.0).

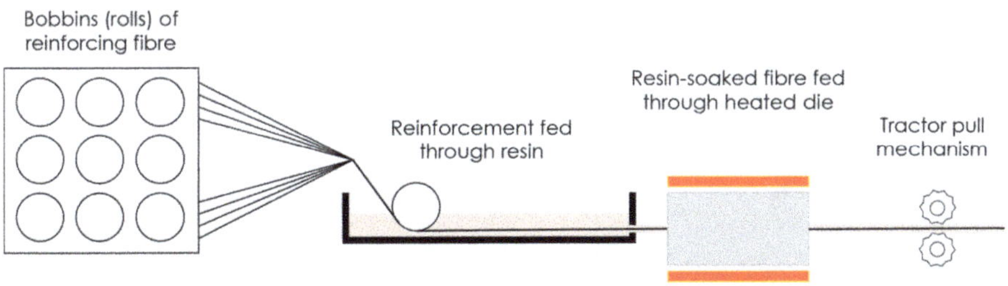

Figure 16.6. Schematic of the general pultrusion process for fibre reinforced plastics.

The resin-soaked fibres then exit the resin tank and are fed into and through a heated die where the composite part gains its shape. The part is pulled out of the die continuously by tractor-type pulling mechanisms and beyond this point, the pultruded part cools down and is cut at a desired length.

While the cross sections are limiting (since they do not change), the pultrusion process offers several advantages. These include cost effectiveness, relatively fast processing times with pultruded lengths continuously produced, fibres tend to align with the direction of pull in a pultrusion process such that the fibre alignment of the final part is usually excellent, and the resin tank can be designed as a closed system thus decreasing the volume of volatile organic compounds emitted from the process.

16.8 Pyroelectric composites

16.8.1 Fundamental description

Pyroelectric materials fall into the larger group of materials known as dielectrics. Dielectrics are electrically insulating materials that can be polarised by an applied electric field. This group comprises piezoelectrics, pyroelectrics and ferroelectrics as shown in figure 16.7. Pyroelectric materials are essentially materials that generate an electrical charge when heated or cooled. This occurs due to a noncentrosymmetrical crystal structure that allows for temperature-activated atomic shifts, resulting in polarisation. While all pyroelectric materials are piezoelectric, and all ferroelectric materials are both pyroelectric and piezoelectric, not all piezoelectric materials are pyroelectric, nor are all pyroelectric materials ferroelectric. Since pyroelectric materials have a non-zero coefficient of thermal expansion, there is a secondary effect on heating, which is mechanical stress and this induces a secondary piezoelectric effect in pyroelectric materials, though the electric charge generation due to temperature changes is predominant.

16.8.2 Heckmann diagram

The relationships existing between the mechanical, electrical and thermal properties in pyroelectric and piezoelectric materials can be understood from a Heckmann diagram, figure 16.8.

16.8.3 Pyroelectric coefficient

The coupling between the electrical polarisation, P, and temperature, T is fundamentally understood through the generalised pyroelectric coefficient, p_i, equation (16.9).

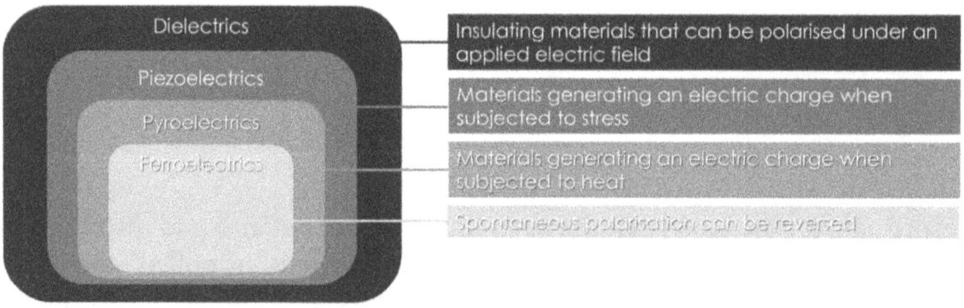

Figure 16.7. Dielectric materials and their sub-groups.

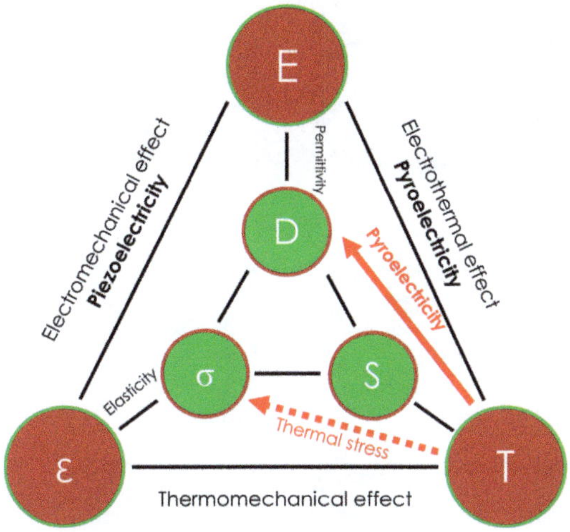

Figure 16.8. Relationships between mechanical, thermal and electrical properties: ϵ, strain; σ, stress; S, entropy; T, temperature; D, displacement; E, electric field.

$$p_i = \frac{\partial P}{\partial T} \tag{16.9}$$

The pyroelectric coefficient can be useful in calculating the current, I, in a pyroelectric materials capacitor where its capacitance varies as a function of temperature, equation (16.10). Here, A refers to the electrode surface area, and $\frac{dT}{dt}$ refers to the rate of change of the temperature [11].

$$I = p_i A\left(\frac{dT}{dt}\right) \tag{16.10}$$

16.8.4 Connectivity of active-passive phases in pyroelectric composites

Pyroelectric composites are combinations of pyroelectric materials with polymerics. The pyroelectric material is termed an active phase (m) while the polymer is termed a passive phase (n). The relationship between active and passive phases ($m - n$) is called connectivity. Connectivity is a simple way of describing the connections between active and passive phases through the composite, as importantly, this will influence the way in which a pyroelectric composite will directionalise the output of electrical energy. A single phase connected in three orthogonal axes (x, y and z) is termed '3'. A single phase connected in only one axis is termed '1'. If connectivity ($m - n$) is then, 0-0, it means that neither the active nor passive phases are connected, while 1-3 means that the active phase is connected in one axis, while the passive phase is connected in 3 axes. In a two-phase (diphasic) pyroelectric composite, there are 10 possible variations, similarly to in piezoelectric composites [2], 0-0, 1-0, 2-0, 3-0, 1-1, 2-1, 3-1, 2-2, 3-2, 3-2, 3-3, 3-3, note that the repetition of 3-2 and 3-3 refers to two different configurations in each hence 12 variations are in fact 10, figure 16.9.

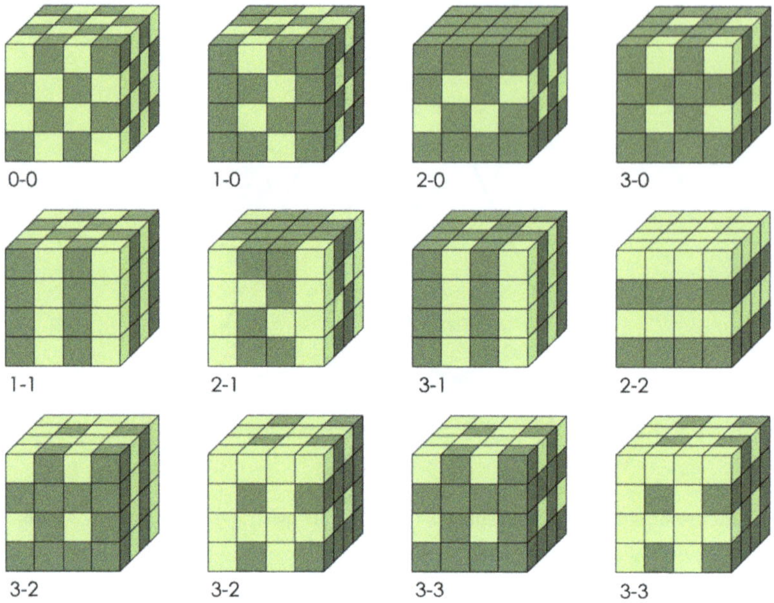

Figure 16.9. 10 variants of connectivity in two-phase pyroelectric composites (note that 3-2 and 3-3 exist in two configurations and thus there are 10 variants rather than 12).

References

[1] Riccardo D 2009 Metal matrix composites: structure and technologies *PhD Thesis* University of Rome Tor Vergata

[2] Uchino K 2017 Piezoelectric composite materials *Advanced Piezoelectric Materials (Second Edition): Science and Technology* 2nd ed (Duxford: Woodhead) pp 353–82

[3] Naslain R R and Pomeroy M R 2001 Ceramic matrix composites: matrices and processing *Encyclopedia of Materials: Science and Technology* pp 1060–6

[4] Low I M 2014 *Advances in Ceramic Matrix Composites* (Woodhead: Cambridge)

[5] Alam P 2011 Porous particle-polymer composites advances in composite materials *Analysis of Natural and Man-Made Materials* ed P Tesinova (Rijeka: InTech) pp 29–54

[6] McAdam G 1951 Some relations of powder characteristics to the elastic modulus and shrinkage of sintered ferrous compacts *J. Iron Steel Inst.* **168** 346–58

[7] Bert C 1985 Prediction of elastic moduli of solids with oriented porosity *J. Mater. Sci.* **20** 2220–4

[8] Alam P 2010 A mixtures' model for porous particle-polymer composites *Mech. Res. Commun.* **37** 389–93

[9] Verpoest I 2000 Composite preforming techniques *Comprehensive Composite Materials* vol 2 (Oxford: Pergamon) pp 623–69

[10] Ribeiro M C S, Fiuza A, Ferreira A, Dinis M D L, Meira Castro A C, Meixedo J P and Alvim M R 2016 Recycling approach towards sustainability advance of composite materials' industry *Recycling* **1** 178–93

[11] Srinivasan M R 1984 Pyroelectric materials *Bull. Mater. Sci.* **6** 317–25

IOP Publishing

Composites Engineering: An A–Z Guide

Parvez Alam

Chapter 17

'Q'

17.1 Quadraxial non-crimp fabric

Quadraxial non-crimp fabrics are part of a larger family of multi-axial fabrics with non-crimp properties. The reason for why these fabrics are non-crimping is because each layer consists of continuous parallel structural fibres, connected by non-structural stitching. While engineering fibres such as glass, carbon or aramid, may be used as the structural fibres, the stitching fibres are often made from polyester as it is lower in cost and has good fibre binding properties. This set-up removes the need for weaving and hence removes the crimp from the fabric. Reduced crimp results in enhanced mechanical performance as the fibres are straighter. Individual fabric layers can be stacked to create fabrics with structural fibres oriented at different angles. Quadraxial fabrics are typically made using [+45/0/−45/90] orientations. A benefit of non-crimping methods when constructing quadraxial fabrics, is that thick fabrics can be made, and with multiple fibre orientations, reducing therefore the need for extensive stacking to build a laminate sequence.

Both weave and stitch methods and simultaneous stitch methods are used to bind unidirectional structural fibres. Using the weave and stitch method, layers of fabric are initially stitched together by skewing and weaving the unidirectional weft for the individual [+45/−45] layers. The [0/90] layers are stitched using convention methods as a single layer composed of orthogonally oriented weft and offer labour saving benefits, and the four layers [−45/0/90/+45], are thence joined together as three layers, with the [0/90] layer sandwiched between the [−45] and the [+45] layers. Special machines are used to create quadraxial fabric by simultaneous stitching. In this method, all layers of the quadraxial are drawn together and then simultaneously stitched through the entire four layers of the fabric.

Non-crimp quadraxial fabrics do not have to be composed of 45° layers of fabric laid on top of one another. Both the layering and the orientations can be varied according to design requirements, manufacturing requirements, and its intended final application. The weight fractions (gsm) are also often varied between the different layers according to requirement, and the stitching fibres might occupy a 0.04–0.08 weight fraction of the entire composite, depending on the stitch tension and the number of fibres stitched. Increasing the stitch tension of a non-crimp quadraxial fabric brings the individual fibres closer together. While this might benefit the mechanical properties of the fabric (due to the development of a higher fibre fraction per unit area), there is a coupled decrease in permeability through the fabric [1], which in turn can affect the effectiveness of fibre wetting and adhesion. Impact energy absorption is one of the strong-points of non-crimp quadraxial fabric as they exhibit superior impact damage tolerance when compared against woven fabrics, when each is compared for use as polymer reinforcing. The optimal orientation variation between layers is typically understood to be 45°.

17.2 Quasi-isotropic laminate

In composite laminate theory (CLT), ply stiffness can be assumed to be under a state of plane stress, equation (17.1), where Q_{ij} is termed the reduced stiffness, and σ_1 and σ_2 are normal stresses, σ_6 is shear stress, ε_1 and ε_2 are normal strains, and ε_6 is shear strain. Q_{ij} is related to stiffness, C_{ij} as $Q_{ij} = C_{ij} - \frac{C_{i3}C_{j3}}{C_{33}}(i, j = 1, 2, 6)$. σ for a 0° lamina can be calculated thence as $\sigma_i = Q_{ij}\varepsilon_j$.

$$\begin{Bmatrix} \sigma_1 \\ \sigma_2 \\ \sigma_6 \end{Bmatrix} = \begin{bmatrix} Q_{11} & Q_{12} & 0 \\ Q_{21} & Q_{22} & 0 \\ 0 & 0 & Q_{33} \end{bmatrix} \begin{Bmatrix} \varepsilon_1 \\ \varepsilon_2 \\ \varepsilon_6 \end{Bmatrix} \tag{17.1}$$

Q_{ij} relates to engineering constants as: $Q_{11} = \frac{E_{11}}{1 - \nu_{12}\nu_{21}}$; $Q_{12} = \nu_{21}Q_{11}$; $Q_{22} = \frac{E_{22}}{1 - \nu_{12}\nu_{21}}$; $Q_{16} = 0$; $Q_{26} = 0$; $Q_{66} - G_{12}$, where E is elastic modulus, ν is Poisson's ratio, and G is shear modulus.

Ply angles in laminates are often varied. Calculating the reduced stiffness so as to take into account the ply angle, $\bar{Q}_{ij}$, requires consideration of the effects of these angles on the properties, equation (17.2), where $m = \cos(\theta)$ and $n = \sin(\theta)$, θ being the ply orientation angle.

$$\begin{Bmatrix} \bar{Q}_{11} \\ \bar{Q}_{22} \\ \bar{Q}_{12} \\ \bar{Q}_{66} \\ \bar{Q}_{16} \\ \bar{Q}_{26} \end{Bmatrix} = \begin{bmatrix} m^4 & n^4 & 2m^2n^2 & 4m^2n^2 \\ n^4 & m^4 & 2m^2n^2 & 4m^2n^2 \\ m^2n^2 & m^2n^2 & m^4 + n^4 & -4m^2n^2 \\ m^2n^2 & m^2n^2 & -2m^2n^2 & (m^2 - n^2)^2 \\ mn^3 & -mn^3 & n^3m - m^3n & 2(n^3m - m^3n) \\ mn^3 & -m^3n & m^3n - n^3m & 2(m^3n - n^3m) \end{bmatrix} \begin{Bmatrix} Q_{11} \\ Q_{22} \\ Q_{12} \\ Q_{66} \end{Bmatrix} \tag{17.2}$$

$\bar{Q}_{ij}$ can then be used to determine the extensional stiffness (A_{ij}), equation (17.3), the coupling stiffness (B_{ij}), equation (17.5), and the bending stiffness (D_{ij}), equation (17.4), where t_k is the thickness of the kth lamina and $\bar{z}_k$ is the distance from the mid-plan to the centroid of the kth lamina.

$$A_{ij} = \sum_{k=1}^{n} (\bar{Q}_{ij})_k t_k \tag{17.3}$$

$$B_{ij} = \frac{1}{2} \sum_{k=1}^{n} (\bar{Q}_{ij})_k t_k \bar{z}_k \tag{17.4}$$

$$D_{ij} = \frac{1}{3} \sum_{k=1}^{n} (\bar{Q}_{ij})_k \left(t_k \bar{z}_k^{-2} + \frac{t_k^3}{12} \right) \tag{17.5}$$

These in turn can then be used to calculate the forces, N, and moments, M, according to equation (17.6), which is shown in expanded form in equation (17.7)

$$\begin{Bmatrix} N \\ M \end{Bmatrix} = \begin{bmatrix} A & B \\ B & D \end{bmatrix} \begin{Bmatrix} \varepsilon^0 \\ \kappa \end{Bmatrix} \tag{17.6}$$

$$\begin{Bmatrix} N_{xx} \\ N_{yy} \\ N_{xy} \\ M_{xx} \\ M_{yy} \\ M_{xy} \end{Bmatrix} = \begin{bmatrix} A_{11} & A_{12} & A_{16} & B_{11} & B_{12} & B_{16} \\ A_{12} & A_{22} & A_{26} & B_{12} & B_{22} & B_{26} \\ A_{16} & A_{26} & A_{66} & B_{16} & B_{26} & B_{66} \\ B_{11} & B_{12} & B_{16} & D_{11} & D_{12} & D_{16} \\ B_{12} & B_{22} & B_{26} & D_{12} & D_{22} & D_{26} \\ B_{16} & B_{26} & B_{66} & D_{16} & D_{26} & D_{66} \end{bmatrix} \begin{Bmatrix} \varepsilon_{xx}^0 \\ \varepsilon_{yy}^0 \\ \varepsilon_{xy}^0 \\ \kappa_{xx} \\ \kappa_{yy} \\ \kappa_{xy} \end{Bmatrix} \tag{17.7}$$

Here, κ_x, κ_y, and κ_{xy} are the laminate curvatures calculated as $\frac{1}{r}$, where r is the radius of curvature. **A quasi-isotropic laminate** is a laminate that exhibits in-plane isotropic material behaviour. The term 'quasi' is used because although the extensional stiffness matrix $[A]$ exhibits isotropic behaviour, the couping stiffness matrix and the bending stiffness matrix, B and D, respectively, may not behave isotropically. As such for a quasi-isotropic laminate, $A_{11} = A_{22}$, $A_{16} = A_{26}$, and $A_{66} = \frac{(A_{11} - A_{12})}{2}$. These are furthermore independent of the laminate angle of rotation.

17.3 Quench hardening (metal matrix composites)

Quenching is a metallurgical method for preserving desired material properties of a particular metallic phase, whilst circumventing undesirable low temperature processes such as phase transformations. Quenching involves the rapid cooling of a metal that has been heated to a desired phase and the process typically takes place in water (containing salt and soda), oil or air/gases. Thermodynamic forces can change the microstructure of a metal and this opportunity becomes available when the rates of quenching are sufficiently slow to allow them to take effect. The rate of quench can be controlled by various factors, one of which is the medium used to cool the metal.

Oil for example, has a lower quenching rate than water (caustics). Quenching is also used to successfully enable other metal-alloy processes such as precipitation hardening, where metals are heated to temperature, treated with a solution and then rapidly quenched to form supersaturated solid solutions containing the different alloying elements. Structural alloys including aluminium, titanium, nickel titanium and steels/stainless steels are increased in yield strength and malleability through (combined) processes such as these.

In metal matrix composites (MMCs), there are four primary strengthening/hardening processes: (1) Orowan strengthening (2) grain/substructure strengthening (3) work hardening and (4) quench hardening. **Quench hardening** is a way of strengthening the composite by enabling the creation of dislocations. The dislocations arise in the composite as a result of a mismatch in the coefficient of thermal expansion (CTE, or α_{CTE}) between the reinforcement and the matrix. Thermal strains, $\varepsilon_{\mathrm{thermal}}$ are ordinarily a product of the CTE and a local variation in temperature, ΔT. When there is a mismatch between the CTE of reinforcement and matrix, the variation in CTE is also taken into account, equation (17.8), where it should be noted that $\varepsilon_{\mathrm{thermal}}$ is also $\varepsilon_{\mathrm{thermal}} = \varepsilon_{\mathrm{total}} - \varepsilon_{\mathrm{mechanical}}$, and $\varepsilon_{\mathrm{total}}$ is the total strain while $\varepsilon_{\mathrm{mechanical}}$ is mechanical strain [2].

$$\varepsilon_{\mathrm{thermal}} = \Delta\alpha_{\mathrm{CTE}}\Delta T \tag{17.8}$$

The total stress in a thermo-mechanically strained body therefore, ordinarily represented by equation (17.9), is then represented by equation (17.10), where E is a local Young's modulus in the direction of loading.

$$\sigma = E\varepsilon - E\alpha_{\mathrm{CTE}}\Delta T \tag{17.9}$$

$$\sigma = E\varepsilon - E\Delta\alpha_{\mathrm{CTE}}\Delta T = E\varepsilon - E\varepsilon_{\mathrm{thermal}} \tag{17.10}$$

Strengthening, i.e. the change in strength, σ, resulting from the CTE mismatch can be expressed by equation (17.11), where γ is a constant reflecting the interactions between dislocations (often simply taken as unity), G_c is the shear modulus of the composite, b is the magnitude of Burgers vector, and ρ_{disloc} is the dislocation density [3].

$$\Delta\sigma = \gamma G_c b\sqrt{\rho_{\mathrm{disloc}}} \tag{17.11}$$

The shear strength of particulate reinforced MMCs, τ_{MMC}, is inversely proportional to interparticulate spacing λ, equation (17.12) and proportional to b, which itself increases as the number of dislocation lines increase in the material. λ can be calculated from the particulate volume fraction V_p and the particulate radius r_p, equation (17.13).

$$\tau_{\mathrm{MMC}} = \frac{Gb}{\lambda} \tag{17.12}$$

$$\lambda = \frac{4(1 - V_p)r_p}{3V_p} \tag{17.13}$$

Since quench hardening increases the number of dislocations due to CTE mismatch, which is itself related to the number of particulates per unit volume, we can expect b to increase through quench hardening, while λ is a pre-existing physical attribute of the MMC being hardened and does not vary through the process of quench hardening.

References

[1] Karanatsis D, James T, Endruweit A and Long A C 2018 Influence of the stitch thread tension on the permeability of carbon fibre non-crimp fabrics *Proceedings of the 18th European Conference on Composite Materials (ECCM18), Athens, Greece, 24th-28th June 2018*

[2] Delannay F 2018 Thermal stresses and thermal expansion in metal matrix composites *Comprehensive Composite Materials II* **4** 213–41

[3] Trojanova Z, Szaraz Z, Lukac P, Drozd Z and Dzugan J 2020 Strengthening and thermally activated processes in an AX61/Saffil metal matrix composite *Crystals* **10** 466

IOP Publishing

Composites Engineering: An A–Z Guide

Parvez Alam

Chapter 18

'R'

18.1 Reaction injection moulding (RIM) techniques in composites engineering

Reaction injection moulding (RIM) is a plastics moulding technique developed in 1969, which has since been adapted for use with both short fibre (reinforced-RIM (RRIM)) and long fibre composites (structural-RIM (SRIM)). A primary difference between RIM and other moulding techniques is that polymerisation and curing reactions only occur after both reactant and polyol (liquid monomers) are injected into the mould as liquid (at low temperatures). Each will be detailed in this section.

18.1.1 Reaction injection moulding (RIM)

The RIM process involves two liquid monomers, a reactant (e.g. isocyanate) and a polyol. Each is combined in a mixing head resulting in a resin, which is injected into a closed mould where it is allowed to polymerise and cure. The process can be conducted at low pressures (0.35–0.5 MPa) and moderate temperatures (60 °C – 70 °C). Air vents are located on the upper side of mould to enable resin to fill the mould. Some of the benefits of RIM include manufacture to tight tolerances, no post-curing, post-machining, or gluing requirements, very fast manufacturing process, and the precise moulding of undercuts. RIM can be used to mould large complex parts such as medical carts, 3D printer parts, door panels for a variety of medical lab instruments, bumpers and crash panels, and more. Figure 18.1(a) shows the general scheme of a RIM process in comparison to RRIM and SRIM processes.

18.1.2 Reinforced reaction injection moulding (RRIM)

In reinforced reaction injection moulding (RRIM), short aspect reinforcements are added to one of the liquid monomers (e.g. the reactant monomer) and this is injected with the polyol liquid monomer, into the closed mould where it polymerises and cures. Common reinforcements include chopped fibres, mica, regrind, etc. The process can be

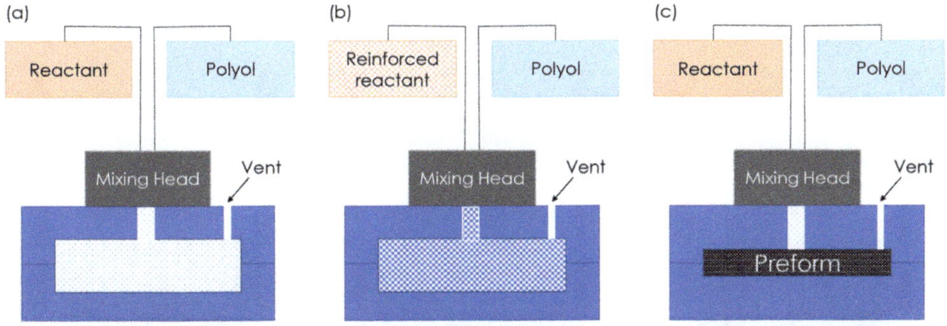

Figure 18.1. Comparative schemes of the different reaction injection processes showing (a) reaction injection moulding (RIM), where only reactant and polyol are injected into the closed mould with an air vent; (b) reinforced reaction injection moulding (RRIM), where short fibre reinforcement is integrated into the reactant and this together with polyol are injected into the closed mould with an air vent; and (c) structural reaction injection moulding (SRIM), where a preform is placed into the closed mould and reactant with polyol are injected into the preform within the closed mould with an air vent.

conducted at low pressures (0.35–0.5 MPa) and moderate temperatures (60 °C–70 °C). Air vents are located on the upper side of mould to enable resin to fill the mould. RRIM shares several of the benefits of RIM, such as manufacture to tight tolerances, no post-curing, post-machining, or gluing requirements, very fast manufacturing process, and the precise moulding of undercuts. Figure 18.1(b) shows the general scheme of a RRIM process in comparison to RIM and SRIM processes.

18.1.3 Structural reaction injection moulding (SRIM)

In structural reaction injection moulding (SRIM), a fibre preform is inserted into the mould prior to injection polymerisation of the liquid monomers. In SRIM neither the reactant nor the polyol are combined with any reinforcement, but these are rather injected in a neat form, similarly to RIM, but directly into the preform, where polymerisation and curing then take place. The SRIM process is conducted at higher pressures and temperatures than similar manufacturing methods such as resin transfer moulding (RTM). While mould pressures in RTM are around 0.3 MPa, they are 2.4MPa in SRIM. Moulding temperatures in RTM are 25 °C–40 °C, whereas in SRIM moulding temperatures are typically 95 °C [1]. Figure 18.1(c) shows the general scheme of a SRIM process in comparison to RIM and RRIM processes. The pressures are higher in SRIM than in RIM and RRIM processes as the SRIM process impregnates a fibrous preform and has a coupled requirement of fibre consolidation if mechanical properties are to be optimised.

18.2 Reactive melt infiltration (RMI) in CMCs

Reactive melt infiltration (RMI) is a liquid phase route to manufacturing ceramic matrix composites (CMCs). The process is only possible if an oxygen-reactive element in the ceramic matrix has (a) a low melting temperature relative to the reinforcement and (b) is able to wet the reinforcement when in a liquid form.

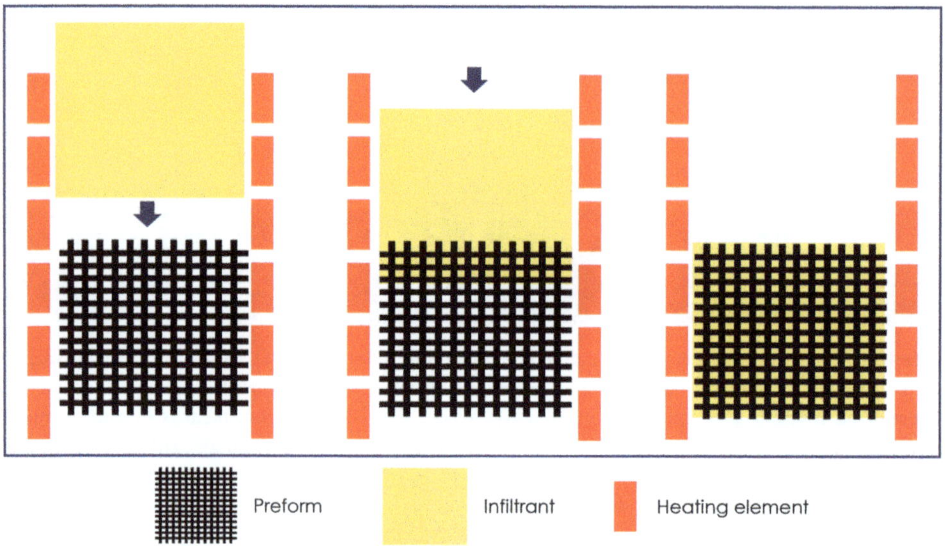

Figure 18.2. Example of an Si infiltrate heated by heating elements and permeating into a C preform (which will form the ceramic matrix) where reactions between the two elements lead to the creation of Si/SiC ceramic matrix composites.

Both aluminium and silicon are candidate elements for RMI since aluminium melts at approximately 650 °C and is reactive with oxygen, while silicon melts at close to 1400 °C and is reactive with carbon. Common final matrix ceramics from these elements are alumina and silicon carbide. Figure 18.2 simplifies the RMI process using Si/SiC CMCs as an exemplar. In this example, a carbon preform (which will form the ceramic matrix) is infiltrated by liquid silicon, which reacts to form silicon carbide (Si + C = SiC). Initially the infiltrate is placed on the preform. The preform itself rests on a boron nitride coated refractory support. The boron nitride coating acts as a release agent for easy release of the CMC. Heating elements are used to melt the infiltrant, in this example silicon, to slightly above 1400 °C or the melting temperature of silicon. This takes place in an inert gas environment to prevent unwanted reactions between silicon and elements other than carbon. The infiltrant enters and reacts with the carbon preform altering its composition to SiC. The SiC retains the shape of the preform and the remainder of the space around the preform is occupied by molten silicon, which once hardened, yields a Si/SiC CMC. The initial reaction of Si with C at the surface of the preform creates a barrier and the kinetics of the reaction is a function of the structure of the carbon used in the preform. The design behind the kinetics and materials involved in the method has received much attention and various recipes are in place for successful RMI manufacturing. Should it be, for example, desirable to incorporate carbon fibres (rather than Si) within an SiC matrix, the procedure first requires that a carbon fibre reinforced polymer is pyrolysed at 1200 °C. At this temperature, the polymer will convert to a porous carbon and the carbon fibres will be physically supported by the presence of the pyrolysed polymer. Once this is complete, a similar RMI process is conducted using

Si as the melt infiltrant, which infiltrates the porous carbon and reacts with it to form the SiC matrix. The SiC matrix surrounds the unconverted carbon fibres, leaving a C/SiC CMC.

Of the different CMC manufacturing routes, cost-performance ratio of RMI is the highest. The resultant CMCs are typically very able to retain mechanical performance in high temperature applications and are dimensionally stable. They are also typically abrasion resistant and corrosion resistant. Example areas of application of RMI manufactured CMCs include as protective armour (silicon carbide particle/silicon) due to their excellent impact energy absorption properties, and in race car and aeroplane brake discs due to their excellent resistance to wear.

18.3 Recycling

Recycling of composite materials is a sensible way of reducing waste and extending the life-cycle use of composite material components. Methods for recycling are diverse and a few of them will be touched upon here. Composite shredding is sometimes a mechanical necessity for transportation, however mechanical methods are also used for separating the components. Crushing is the usual route for mechanically separating components. Smaller material deposits are usually the reinforcement materials and the large deposits are frequently polymer matrix. Mechanical separation nevertheless does not guarantee any level of purity. Thermoplastic matrices can be melted after shredding in large spinning containers however, thermosets require more detailed attention. Here we discuss different recycling techniques for different common types of composites.

18.3.1 Recycling of glass fibre reinforced plastics (GFRP)

It can be economically attractive to recycle at least the fibres from GFRP. Chemical recycling is common practice and acids used to help separate thermoset polymer matrices from glass fibre reinforcements. The polymer effectively dissolves in the acid and can be extracted through chemical addition and drying. The mechanical properties of chemically extracted resin have been shown to, in some cases, be superior to the virgin resin. T-glass/C-glass fibres (alkali-lime glass with high boron oxide content) fair better under acidic treatment as compared to E-glass (alumino-borosilicate glass with less than 1 wt % alkali oxides). Pyrolysis (heat treatment) is another means by which glass fibres can be separated from polymeric matrices. Oil that may form (e.g. when pyrolysing polyester) can be recovered and has properties of fuel oil with a calorific value of ca. $34 MJkg^{-1}$. The main gases that evolve through pyrolysis of polystyrene are CO and CO_2. Glass fibres recovered from pyrolysis can be successfully re-used as the reinforcing component in reinforced glass fibre GFRPs. The most recyclable glass fibre is A-glass.

18.3.2 Recycling of carbon fibre reinforced plastics (CFRP)

Carbon fibres are more expensive than glass fibres (ca. 10:1 price differential) and have superior properties. Recycling carbon fibres therefore, is economically more rewarding than for glass fibre. Various pyrolysis techniques are currently in use. Vacuum pyrolysis operates at 500 °C and resin can be recovered and re-used. The fibres at this temperature however retain some level of oxidation or char. Microwave pyrolysis, which exploits the conductive properties of carbon, results in faster resin decomposition and omits the problems of fibre charring. Chemical treatments are also used to separate carbon fibre and matrix materials. Similarly to GFRPs, acids (often nitric acid ca. 8–14 M l^{-1}) can be used to dissolve away polymeric matrix material. Organic solvents such as tetralin (1,2,3,4-tetrahydronaphthalene; $C_{10}H_{12}$) can also be used at high pressures (ca. 2 MPa) and temperatures (ca. 350 °C) under prolonged time-scales (ca. 2 h). The supercritical process is another method for separating composite phases. This refers to, in this case, critical pressures and temperatures at which specific liquids will best degrade matrix material. As an example, epoxy in CFRP can be degraded in ethanol 120 ml at temperatures of 255 °C and pressures of 7.6 MPa. There is little loss in the tensile strength of the carbon fibre after extraction. Acids show the lowest losses (ca. 3% from the original strength), organic solvents at ca. 4% and pyrolysis in oxygen results in a strength loss of slightly over 4%. Recycling of the decomposed matrix is only possible with the acid treatment. Comparing between pyrolysis and acid techniques shows that pyrolysis emits the highest levels of CO_2 and methane, ca. 5-fold.

18.3.3 Recycling of metal matrix composites (MMC)

Metal matrix composites are usually recycled by means of separating components at the bonded interfaces. Techniques in use include the melting of composite scrap followed by recasting, salt flux for ceramic particle-metal matrix composites, hot pressing followed by extrusion, and electrolysis in recyclable ionic liquids (yielding high purity aluminium deposits). Salt flux is a method of de-wetting ceramic particles using molten salts. In this method, molten salts wet ceramic particles in the presence of molten aluminium allowing for the physical removal of particles.

18.3.4 Recycling of reinforced concrete

Concrete is easily recycled. Crushing it allows for the recovery of aggregates that can be re-used. Crushed concrete debris can also be used as roadbed material and as such, expensive recycling processes need not be used.

18.3.5 Recycling of wood-based composites

As a general rule, wood composites comprised of larger dimension woody components can be recycled into composites comprising smaller dimension woody components. There are three main methods that are used to separate the woody components of wood-based composites. The first is mechanical disintegration, a dry process where woody components are mechanically crushed into smaller parts that can be re-used. A process such as this results in blocky woody components with considerable levels of damage. One point of note with mechanical disintegration, is that if the wood is infused with resin or in contact with resin, then the final product will include the original resin, which may need further separation. This method of disintegration is generally seen as limiting to the end-use of the recycled wood-composite. The Reholz-process is an improvement on the mechanical disintegration method as it slices the composite into flat discs, rather than into blocky chunks. A second method involves thermo-hydrolytic disintegration. Here, steam and pressure are used to separate hydrolysable resins from woody reinforcements. With certain wood composites such as panel-boards, the boards are pre-crushed prior to thermo-hydrolytic disintegration as it can be helpful in developing a hygroscopic equilibrium enabling better control therefore, of dimensional changes that may arise through swelling. The final general method for recycling the woody component in wood composites is chemical pulping. The benefit of this process is that it is not restricted to only hydrolysable resins [2]. The resulting recycled material from this process is wood pulp, which can be re-used in the manufacture of particle and fibre boards, paper-boards and paper.

18.4 Reinforcement

Reinforcements are materials combined with a matrix to improve mechanical performance. These effectively, reinforce the matrix material, even though they can sometimes exceed the matrix in volume fraction. Table 18.1 provides examples of categories and types of reinforcement, and provides examples of materials used in each.

18.5 Resin transfer moulding (RTM)

Resin transfer moulding (RTM) is an out-of-autoclave (OoA) low pressure, closed moulding process where a resin/curing agent mix is fed into a closed mould containing a preform or a fibrous packing (oriented or random). Whether heat is applied to the mould or not will depend on the curing process for the catalysed resin. Once cured, the male and female segments of the mould can be separated and the composite part is complete. Gel coat is conventionally applied to the mould prior to moulding. RTM does not have any resin flow assistance from a vacuum (such as in the vacuum assisted resin transfer moulding (VARTM) process) but rather, displaced air is released via mould vents. For this to be possible, the injection pressure of the catalysed resin must be higher than the air pressure in the operating environment. Figure 18.3 shows a generalised schematic of the RTM process.

Due to the simplicity of the process, RTM can be used with a wide variety of resins and fibres/preforms, including toughened resin systems containing particulate inclusions. Parts made through the RTM process will typically have a good quality surface finish and this is due to the use of matched metallic male and female moulding tools. The process is generally very quick and a relatively high OoA production volume is possible provided suitably fast curing resins are used. High performance parts with high reinforcement volume fractions (ca. 65%) can be made through the application of pressure, though these parts are smaller than in VARTM processes, which can be used for the production of very large parts. Generally, larger parts require higher moulding clamping pressures and the clamping pressure is thus a limiting factor in production of larger part sizes. Due to the size difference in the

Table 18.1. Different categories and types of reinforcement, and common materials used for each.

Category	Type	Example materials
Structural	Rebars	Steel
	Plates/panels	Steel/Aluminium/Fibre reinforced plastic/ Wood–plastic composites
	Mesh	Steel/Iron/Glass fibre
	Rods	Steel/Fibre reinforced plastic
	Fibres	Natural fibres/Glass fibres/Carbon fibres
	Mats	High density polyethylene/Wood strand based/Glass strand
	Coarse aggregates	Gravel/Pebbles/Cobbles/Ballast
	Fine aggregates	Sand/Crushed stones/Slag
Fibre	Continuous (long) fibre	Carbon/Glass/Boron/Nylon/Natural fibre
	Discontinuous (short) fibre	Carbon/Glass/Boron/Nylon/Natural fibre
	Chopped strand	Glass/Natural fibre
	Fibre bundle	Natural fibre bundle
	Woven fibre	Carbon/Glass/Polymer
	Knitted fibre	Natural fibre/Synthetic polymer
	Nanofibre	Cellulose/Carbon/Polymer
	Whisker	Cellulose/Alumina/Silicon carbide/Iron/ Hyroxyapatite/Iron/Silicon
Particle	Low aspect particle	Ground calcium carbonate/Precipitated calcium carbonate/Talcs/Crystalline silica/ Carbon black/Silicon carbide/Alumina/ Zirconia
	High aspect particle	Kaolin/Bentonite/Mica
	Porous particle	Bentonite/Templated calcium carbonates/ Templated calcium phosphates
	Nanoparticle	Carbon nanotube/Cellulose/Chitin/Starch

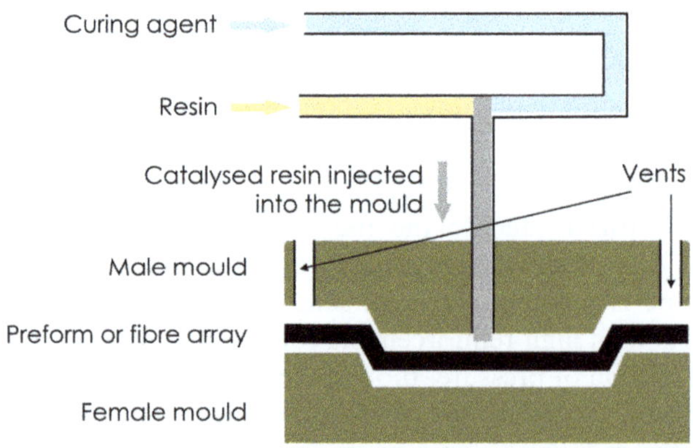

Figure 18.3. Generalised schematic of the resin transfer moulding (RTM) process.

parts between RTM and VARTM, RTM has lower quantities of air bubble voids. Manufacturing to close dimensional tolerances is enabled through the RTM process and reinforcement material can be hand-laid to suit such that combinations of preforms or fibre combinations may be included within the same part to produce hybrid composites. Since moulding is possible to near-net shape, the process experiences low levels of material waste. This said, undercut geometries can still be difficult to mould. Important process design parameters in RTM include: preform permeability together with resin viscosity, mould temperature, vent control, injection pressure, and mould shape/configuration.

18.6 Reinforced concrete

Reinforced concrete is an important construction material used widely in large scale construction. The concrete itself is made up of water, cement, fly ash, coarse aggregates, fine aggregates and plasticisers. Coarse aggregates make up the highest weight per unit volume in the mixture followed by fine aggregates, fly ash, cement, water and plasticisers in that order. A benefit of concrete is that silica and other solid wastes from, e.g. metals refining industries, can be used as aggregate. Concrete is nevertheless very weak in tension and bending. While in compression its failure strength is close to 32 MPa, in tension it is only 2 MPa and in bending only 6 MPa. As such concrete is commonly reinforced to improve its strength and behaviour in tension, and to enable an albeit small, brittle to ductile transition in its tensile or flexural failure mode. This section aims to introduce the reader to the different types of reinforcing material used in concrete and the ways in which they are applied.

18.6.1 Steel reinforcements in concrete

Mild steels are by far the most common material used to reinforce concrete. Reinforcements are in the form of deformed steel bars (steel rebars), threaded steel bars, cold drawn wire, welded wire fabric and deformed welded wire fabric. Some of these are covered in more detail below [3].

- **Deformed steel bars (rebars):** Steel rebars, figures 18.4(a) and (b), are the most common style of concrete reinforcement. They are effective not only because steel has excellent tensile properties and high ductility, but also because they have deformed surfaces which enable mechanical interlocking at the interface between the reinforcement and the concrete matrix. Rebar sizes and materials (billet steel, low alloy steel and rail and axle steel) are provided in ASTM A 615M/A 615, ASTM A 706M/A 706, and ASTM A 996M/A 996, respectively [4–6] and the minimum allowable yield strengths of steel rebars are 300–520 MPa.
- **Threaded steel bars:** Threaded steel bars, figure 18.4(c), have essentially a screw thread machined into the surface which creates a more systematic and equally spaced protuberant surface than steel rebars. The protuberant surface enables mechanical interlocking between the reinforcement and the concrete matrix. Threaded steel bars can be spliced, sometimes using couplings or by anchoring them through steel plates. This ensures that finite length threaded

Figure 18.4. Examples of (a) a deformed steel bar, (b) laying of deformed steel bars, (c) tensile fractured threaded steel bars including a coupling, and (d) welded wire fabric. Reprinted from (a) [7], (b) photo by Ricardo Gomez Angelon Unsplash, published on 24 February 2020 CC-BY, (c) [8], (d) [9] with permission from MDPI (CC BY 4.0)

steel bars still provided a continuous reinforcement–matrix bond at any length of concrete.

- **Welded wire fabric:** Welded wire fabric, figure 18.4(d), is a welded mesh of steel wires resulting in square or rectangular repeating units. It is particularly important as a reinforcement as it resists temperature and/or shrinkage fracture in concrete slabs. It may be manufactured using plain or deformed wire. Relevant standards for welded wire fabric are ASTM A884/A884M-19e1 [10] which covers deformed steel bar mats, ASTM A1064/A1064M [11], which covers carbon-steel wire and welded wire reinforcement, plain and deformed, and ASTM A884M/A884 [12] which covers epoxy-coated steel wire and welded wire fabrics.

The corrosion of steel results in volumetric changes in the reinforcement, which can lead to micro and subsequently macro cracking, and which weakens the bond at the steel-concrete interface. Since steel corrodes under various conditions, a variety of treatments may be applied to decrease the rate or likelihood of corrosion. These may include the use of epoxy coatings, galvanising, and the use of stainless steel reinforcements.

18.6.2 Fibre reinforced plastic reinforcements in concrete

Fibre reinforced plastic (FRP) composites are lightweight alternatives to steel that also offer superior corrosion resistance. They are nevertheless more susceptible to moisture ingress and have the disadvantage of failing abruptly and catastrophically on reaching a maximum strength. Steel in comparison is highly ductile and will hold load, absorbing extra mechanical energy, prior to catastrophically failing.

Embedding the FRP into concrete can protect the reinforcement from moisture ingress. The main fibre types used in FRP reinforcements for concrete are glass, carbon and aramid. Of the three, carbon is the least reactive and most chemically resistant. FRP reinforced concrete is often considered only suitable as a tension reinforcement.

18.6.3 Fibre reinforcements in concrete

Fibre reinforcements can be incorporated directly into the concrete matrix to form fibre reinforced concrete (FRC), which improves durability. Common reinforcing fibres used include steel, glass, nylon, polyester, polypropylene, and natural fibres. The incorporation of fibre reinforcements directly into the concrete matrix has the advantage of decreased shrinkage related cracking, improved impact damage tolerance, and heightened energy absorption. Fibres commonly sized between 10 and 75 mm, are added to the concrete during the batching operation. They can also if needed be added to the concrete on site. While steel fibres are added to the rotating mixer, nylon, polyester and polypropylene fibres are added to the mixing system in a pre-weighed degradable bag. Once in the mix, the bag degrades, releasing the fibres into the concrete as it mixes. Steel fibres can be from cut wire, cold drawn wire, or melt extracted (ASTM A 820 [13]) and are made from either carbon steel or stainless steel. FRCs are generally used in slab-type construction elements but FRC can also be used to repair fracture damaged concrete.

18.7 Rule of mixtures (ROM)

The rule of mixtures (ROM) is essentially a weighted mean of properties of individual phases or components within a composite. The rule has been modified to suit a variety of different composite structures, loading conditions, reinforcement types, and sizes, and to predict different composite properties. For this section we will focus on only the simple (classical) ROM models, as other more specialised ROM models have been detailed in other sections of the book.

18.7.1 Elastic modulus of continuous fibre unidirectional composites loaded in the fibre axis

For continuous fibre unidirectional composites loaded in the fibre axis (a Voigt model), the composite elastic modulus, E_c, can be calculated according to the ROM shown in equation (18.1), where E_f and E_m refer to the moduli of the fibre and matrix, respectively, and V_f and V_m refer to the volume fractions (expressed as a real number between 0 and 1) of the fibre and matrix, respectively.

$$E_c = E_f V_f + E_m V_m \tag{18.1}$$

This ROM model assumes that both fibre and matrix phases are subjected to equal strain, that there is perfect bonding between the fibre and matrix phases, and that the fibres are aligned perfectly parallel to the loading axis, are continuous and uniformly distributed.

18.7.2 Elastic modulus of continuous fibre unidirectional composites loaded perpendicular to the fibre axis

For continuous fibre unidirectional composites loaded perpendicular to the fibre axis (a Reuss model), the composite elastic modulus, E_c, can be calculated according to the ROM shown in equation (18.2), where E_f and E_m refer to the moduli of the fibre and matrix, respectively, and V_f and V_m refer to the volume fractions (expressed as a real number between 0 and 1) of the fibre and matrix phases, respectively. This ROM is commonly referred to as either the transverse rule of mixtures, or the inverse rule of mixtures.

$$E_c = \left(\frac{V_f}{E_f} + \frac{V_m}{E_m} \right)^{-1} \tag{18.2}$$

This ROM model assumes that both fibre and matrix phases are subjected to equal stress, that there is perfect bonding between the fibre and matrix phases, and that the fibres are aligned perfectly parallel perpendicularly to the loading axis, and are continuous and uniformly distributed. The Halpin–Tsai model (described in section 8.1) is another ROM model that can be used for particle reinforced composites.

18.7.3 Elastic modulus of short fibre composites

The modulus of short fibre composites is complex due to factors such as fibre size distribution and fibre orientation. The shear lag model (also Cox's model) is a classical model used to determine the elastic modulus of short fibre composites and is shown in equation (18.3), where E_f and E_m refer to the moduli of the fibre and matrix, respectively, and V_f and V_m refer to the volume fractions (expressed as a real number between 0 and 1) of the fibre and matrix phases, respectively, L is the average fibre length, and η is a shear lag parameter shown in equation (18.4), where r refers to the average fibre radius, ν_m is the Poisson's ratio of the matrix, and P_f is a packing parameter which for square section packings is $P_f = \pi$ and for hexagonal packings is $P_f = \frac{2\pi}{\sqrt{3}}$.

$$E_c = E_f \left(1 - \frac{\tanh\left(\frac{\eta L}{2}\right)}{\left(\frac{\eta L}{2}\right)} \right) V_f + E_m V_m \tag{18.3}$$

$$\eta = \frac{1}{r} \left(\frac{2E_m}{E_f(1 + \nu_m)\ln\left(\frac{P_f}{V_f}\right)} \right)^{\frac{1}{2}} \tag{18.4}$$

18.7.4 Elastic modulus of uniformly distributed particle reinforced non-porous composites

For uniformly (homogeneously) distributed particle reinforced non-porous composites loaded in any direction, the composite elastic modulus, E_c, can be calculated according to the ROM shown in equation (18.5), where E_p and E_m refer to the moduli of the particle and matrix, respectively, and V_p and V_m refer to the volume fractions (expressed as a real number between 0 and 1) of the particle and matrix phases, respectively. This ROM is essentially the same as the transverse or inverse rule of mixtures described above.

$$E_c = \left(\frac{V_p}{E_p} + \frac{V_m}{E_m} \right)^{-1} \tag{18.5}$$

This ROM model assumes that both particle and matrix phases are subjected to equal stress, that there is perfect bonding between the particle and matrix phases, and that the particles are uniformly distributed throughout the matrix. The Halpin–Tsai model (described in section 8.1) is another ROM model that can be used for particle reinforced composites.

18.7.5 Strength of continuous fibre unidirectional composites loaded in the fibre axis

For continuous fibre unidirectional composites loaded in the fibre axis (a Voigt model), the composite strength, σ_c, can be calculated according to the ROM shown in equation (18.6), where σ_f and σ_m refer to the strengths of the fibre and matrix, respectively, and V_f and V_m refer to the volume fractions (expressed as a real number between 0 and 1) of the fibre and matrix, respectively.

$$\sigma_c = \sigma_f V_f + \sigma_m V_m \tag{18.6}$$

Predicting strength using this model is not always accurate. This is because the strength of continuous fibre unidirectional composites also depends on a number of other factors such as the strength of bonding at the fibre–matrix interface, which is not accounted for in this model.

18.7.6 Strength of continuous fibre unidirectional composites loaded perpendicular to the fibre axis

For continuous fibre unidirectional composites loaded perpendicular to the fibre axis (a Reuss model), the composite strength, σ_c, can be calculated according to the ROM shown in equation (18.7), where σ_f and σ_m refer to the strengths of the fibre and matrix, respectively, and V_f and V_m refer to the volume fractions (expressed as a real number between 0 and 1) of the fibre and matrix phases, respectively.

$$\sigma_c = \left(\frac{V_f}{\sigma_f} + \frac{V_m}{\sigma_m} \right)^{-1} \tag{18.7}$$

Predicting strength using this model is not always accurate. This is because the strength of continuous fibre unidirectional composites also depends on a number of other factors such as the strength of bonding at the fibre–matrix interface, which is not accounted for in this model.

18.7.7 Strength of short fibre composites

The strength of short fibre composites is complex due to factors such as fibre size distribution and fibre orientation. The shear lag model (also Cox's model) is a classical model that can be used to determine the strength of short fibre composites and is shown in equation (18.8), where σ_f and σ_m refer to the strengths of the fibre and matrix, respectively, and V_f and V_m refer to the volume fractions (expressed as a real number between 0 and 1) of the fibre and matrix phases, respectively, L is the average fibre length, and η is a shear lag parameter shown in equation (18.9), where r refers to the average fibre radius, E_f and E_m refer to the moduli of the fibre and matrix, respectively, ν_m is the Poisson's ratio of the matrix, and P_f is a packing parameter which for square section packings is $P_f = \pi$ and for hexagonal packings is $P_f = \frac{2\pi}{\sqrt{3}}$.

$$\sigma_c = \sigma_f\left(1 - \frac{\tanh\left(\frac{\eta L}{2}\right)}{\left(\frac{\eta L}{2}\right)}\right)V_f + \sigma_m V_m \tag{18.8}$$

$$\eta = \frac{1}{r}\left(\frac{2E_m}{E_f(1 + \nu_m)\ln\left(\frac{P_f}{V_f}\right)}\right)^{\frac{1}{2}} \tag{18.9}$$

Predicting strength using this model is not always accurate. This is because the strength of short fibre composites also depends on a number of other factors such as the strength of bonding at the fibre–matrix interface, which is not accounted for in this model.

18.7.8 Strength of uniformly distributed particle reinforced non-porous composites

For uniformly (homogeneously) distributed particle reinforced non-porous composites loaded in any direction, the composite strength, σ_c, can be calculated according to the ROM shown in equation (18.10), where σ_p and σ_m refer to the moduli of the particle and matrix, respectively, and V_p and V_m refer to the volume fractions (expressed as a real number between 0 and 1) of the particle and matrix phases, respectively.

$$\sigma_c = \left(\frac{V_p}{\sigma_p} + \frac{V_m}{\sigma_m}\right)^{-1} \tag{18.10}$$

Predicting strength using this model is not always accurate. This is because the strength of particle composites also depends on a number of other factors such as the strength of bonding at the particle–matrix interface, which is not accounted for in this model.

References

[1] Davies G 2003 *Materials for Automobile Bodies* 1st edn (Burlington: Butterworth-Heinemann)

[2] Kharazipour A and Kues U 2007 Recycling of wood composites and solid wood products *Wood Production, Wood Technology, and Biotechnological Impacts* ed U Kues (Netherlands: Universitatsverlag Gottingen)

[3] ACI Education Bulletin E2-00 2000 (Farmington Hills, MI: American Concrete Institute)

[4] ASTM A615/A615M-09 2009 *Standard Specification for Deformed and Plain Carbon-Steel Bars for Concrete Reinforcement* (West Conshohocken, PA: ASTM International)

[5] ASTM A706/A706M-06 2006 *Standard Specification for Low-Alloy Steel Deformed and Plain Bars for Concrete Reinforcement* (West Conshohocken, PA: ASTM International)

[6] ASTM A996/A996M-16 2016 *Standard Specification for Rail-Steel and Axle-Steel Deformed Bars for Concrete Reinforcement* (West Conshohocken, PA: ASTM International)

[7] Di B, Wang J, Li H, Zheng J, Zheng Y and Song G 2019 Investigation of bonding behavior of FRP and steel bars in self-compacting concrete structures using acoustic emission method *Sensors* **19** 159

[8] Cho S, Lim M and Lee C 2021 Study on the bending and joint performances of reinforced concrete beams using high-strength rebars *Sustainability* **13** 3482

[9] Murali G, Amran M, Fediuk R, Vatin N, Raman S N, Maithreyi G and Sumathi A 2020 Structural behavior of fibrous-ferrocement panel subjected to flexural and impact loads *Materials* **13** 5648

[10] ASTM A184/A184M-19 2019 *Standard Specification for Welded Deformed Steel Bar Mats for Concrete Reinforcement* (West Conshohocken, PA: ASTM International)

[11] ASTM A1064/A1064M-18a 2018 *Standard Specification for Carbon-Steel Wire and Welded Wire Reinforcement, Plain and Deformed, for Concrete* (West Conshohocken, PA: ASTM International)

[12] ASTM A884/A884M-19e1 2019 *Standard Specification for Epoxy-Coated Steel Wire and Welded Wire Reinforcement* (West Conshohocken, PA: ASTM International)

[13] ASTM A820/A820M-16 2016 *Standard Specification for Steel Fibers for Fiber-Reinforced Concrete* (West Conshohocken, PA: ASTM International)

IOP Publishing

Composites Engineering: An A–Z Guide

Parvez Alam

Chapter 19

'S'

19.1 Sandwich panels

Sandwich panels are important engineering composites as they have excellent specific properties of strength and stiffness, especially in bending. A sandwich panel is comprised of a core material sandwiched between two outer skin materials. The most usual construction for a sandwich panel comprises two stiff and strong skins and a lightweight core. Example combinations include paper/gypsum/paper plasterboards, FRP/honeycomb/FRP used in a very broad variety of engineering applications, e.g. aeroplane wings, steel/timber/steel composite crossbeams, cardboard/corrugated cardboard/cardboard box panels, aluminium/polyethylene foams/aluminium insulators. Sandwich structures are particularly good for resisting bending and buckling as their stiff layers are situated away from the sandwich structure neutral axis, which increases the effective second area moment of the composite. Figure 19.1 shows the basic construction of a sandwich panel composite.

19.1.1 Elastic beam theory for sandwich panels under three-point bending

Three-point bending is a commonly used and relatively simple way of analysing the bending resistance of sandwich panels. Assuming a simply supported sandwich panel, figure 19.2, with perfect bonding at the skin-core interfaces, and that bends perfectly in the *xz* plane, with no anticlastic bending in the *yz* plane, we can use parallel axis theorem to deduce the flexural rigidity (EI) of the sandwich panel, equation (19.1), where b is the width of the sandwich, t is the thickness of the skin, d is the distance from centroid to centroid of the skins, c is the thickness of the core, and E_{skin} and E_{core} are the Young's moduli of the skin and core materials, respectively. $\frac{E_{\text{skin}}bt^3}{6}$ describes the bending stiffness of the skins, $\frac{E_{\text{core}}bc^3}{12}$ describes the bending stiffness of the skins and the core about the centroid, and $\frac{E_{\text{skin}}btd^2}{2}$ describes the bending stiffness of the skins about the centroid of the sandwich panel beam

doi:10.1088/978-0-7503-3527-0ch19

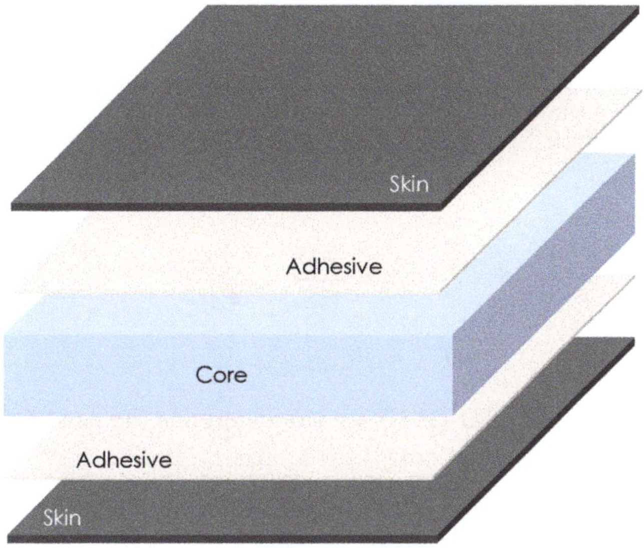

Figure 19.1. Generic lay-up of a sandwich panel comprising two panel skins adhesively bonded to a sandwich core material.

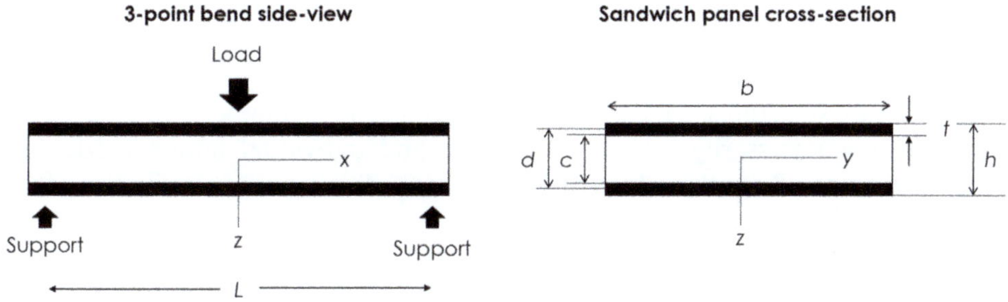

Figure 19.2. Dimensions of three-point bend set up and sandwich panel cross-section.

itself. The first two terms are usually extremely small compared to the third term and the bending of the skins about the centroid of the sandwich panel beam itself dominates. The equation is often simplified to equation (19.2).

$$EI = \frac{E_{\text{skin}}bt^3}{6} + \frac{E_{\text{core}}bc^3}{12} + \frac{E_{\text{skin}}btd^2}{2} \tag{19.1}$$

$$EI = \frac{E_{\text{skin}}btd^2}{2} \tag{19.2}$$

The shear rigidity, AG, can be expressed according to equation (19.3), where G_{core} is the shear modulus of the core and the deflection, δ, under load, F, is the sum of the deflection due to bending and the deflection due to shear, δ_b and δ_s, respectively, equation (19.4). It can be noted that G_{core} is a function of the relative density of the core $\rho_{\text{core},r} = \left(\frac{\rho_{\text{core}}}{\rho_{\text{skin}}}\right)$ and the modulus of the solid material making up the core structure, $E_{\text{core},s}$, equation (19.5), where B and m are fitting parameters that can be obtained from experimental data.

$$AG = \frac{bd^2 G_{\text{core}}}{c} \tag{19.3}$$

$$\delta = \delta_b + \delta_s = \frac{FL^3}{48EI} + \frac{FL}{4AG} \tag{19.4}$$

$$G_{\text{core}} = BE_{\text{skin}}\, \rho_{\text{core},\, r}^{m} \tag{19.5}$$

Collecting equations (19.2)–(19.5), we can determine the bending stiffness of the sandwich panel, $\frac{F}{\delta}$, equation (19.6) [1].

$$\frac{F}{\delta} = \frac{2bd^2}{\frac{L^3}{6tE_{\text{skin}}} + \frac{cL}{BE_{\text{core},s}}\rho_{\text{core},r}^{m}} \tag{19.6}$$

19.1.2 Failure modes of sandwich panels

Being complex structures, sandwich panels may fail in a number of ways. Failure modes are not necessarily mutually exclusive and individual failure modes may trigger the onset of another. Figure 19.3 shows a few common failure modes in sandwich panel composites.

Face yielding, figure 19.3(a), occurs when either, the yield stress in compression of the skin subjected to compression from a bending moment is surpassed, or micro-buckling occurs in the skin which leads to premature yielding in the skin in compression. Core shear, figure 19.3(b), occurs when the shear yield stress is reached before failure occurs in the skins. This tends to happen when the skins are significantly stiffer and thinner than the core. Face wrinkling, figure 19.3(c), is macro-buckling of the compressive face skin of the sandwich panel. Where honeycomb cores are used, this is typically measured by the unit cell dimensions of the honeycomb such that face wrinkling is more than one unit cell dimensions of the honeycomb. Buckling may push inwards into the core material, or outwards from the core material. Core indentation, figure 19.3(d), or local indentation, occurs when a point load causes local indentation of the skin coupled to crushing beneath in the core. The area of the indentation footprint is a function of the stiffness of both skin and core materials.

19.1.3 Relevant standards

Useful standards for sandwich panel testing include the ASTM C393/C393M-20 [2], which is the Standard Test Method for Core Shear Properties of Sandwich Constructions by Beam Flexure, and the ASTM D7249/D7249M-20 [3], which is

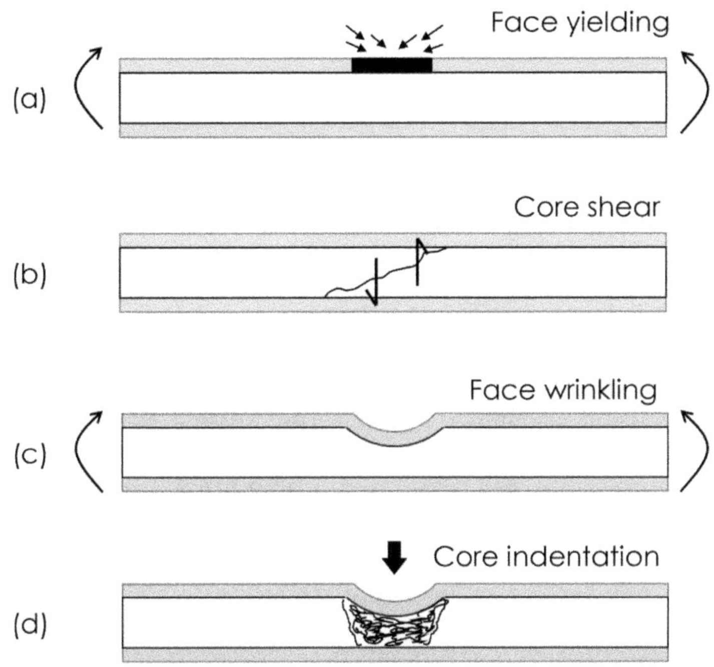

Figure 19.3. Examples of common failure modes in sandwich panels (a) face yielding, (b) core shear, (c) face wrinkling and (d) core indentation.

the Standard Test Method for Facesheet Properties of Sandwich Constructions by Long Beam Flexure.

19.2 Sizing

Sizing is a thin layer of organic material coating the surface of reinforcing fibres such as carbon, basalt and glass. Sizing layers are added to the fibres as part of the manufacturing process and are important coating materials that protect the fibres during handling and processing, while additionally influencing the effectiveness of the fibres within a polymer matrix. The sizing, due to its location at the fibre interface, also affects the properties of the interphase that develops in matrix materials at reinforcement interfaces. Wetting, adhesion, hydrophobicity and chemical resistance are examples of properties that are influenced by the presence of sizing. Sizing of glass fibres can be applied as part of the manufacturing process.

19.2.1 Applicator roll application of sizing to fibre surfaces

The sizing of monofilament glass fibres is normally conducted via an applicator roll after the glass fibres exit the bushings and when the fibres reach a critical temperature. Nevertheless, more complex hybrid fibre systems can also be sized collectively. Figure 19.4 shows an example of this, where fibre sizing is applied at the commingling stage of a hybrid yarn spinning process. The temperature at the time of coating should be below a critical temperature, for glass fibre sizings this is often <100 ° C.

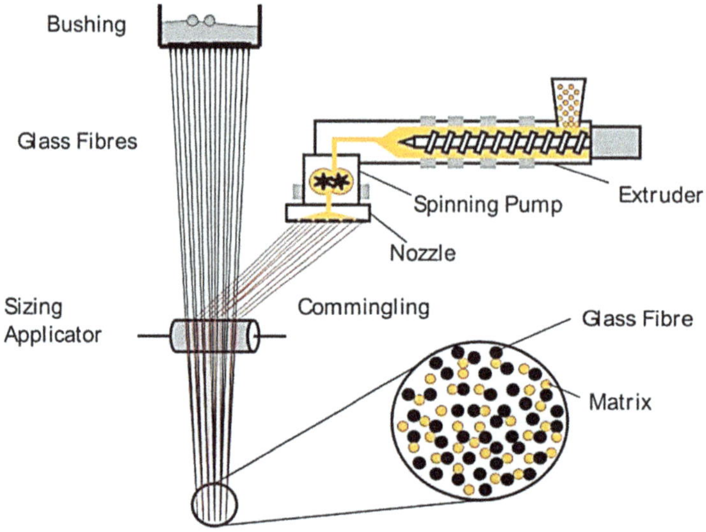

Figure 19.4. Principle of online hybrid yarn spinning showing fibre sizing at the commingling stage of the process. Reprinted from [4] with permission from MDPI (CC BY 4.0).

19.2.2 Fibre specific sizing

When used in engineering composites, sizing aims to create a strong bond between the fibre and the polymer matrix. Epoxy (a thermosetting polymer) is the most common matrix used in fibre reinforced plastic (FRP) and as such, sizing used for glass and basalt fibres contains silane based compounds (organosilane coupling agents) to enable strong bonding between the glass and the epoxy. Organosilane coupling agents make up around 10% by weight of the final sizing. Carbon fibre sizing technologies are less demanding when used with epoxy resins as the carbon fibres are wet well by epoxy and these form relatively strong secondary bonds. Carbon fibre used in epoxy matrices are commonly coated with epoxy-based sizing.

19.2.3 High temperature sizing

In certain high temperature applications, the use of traditional sizing on fibres will most likely result in sizing degradation. The consequence of this is discontinuity at interfaces, microvoids from outgassing and poor adhesion, leading to low mechanical performance and the early onset of composite failure. In high temperature applications, polyimide-based sizing are sometimes used as these have thermal stability up to a few hundred degrees Celsius.

19.3 Slurry impregnation and hot processing of CMCs

Slurry impregnation and hot pressing (SIHP) also known as slurry impregnation and consolidation (SIC), is a common method for manufacturing CMCs. The process has limitations in that only certain reinforcements can be used that show stability under high temperatures (e.g. neat SiC reinforcements). The process begins with

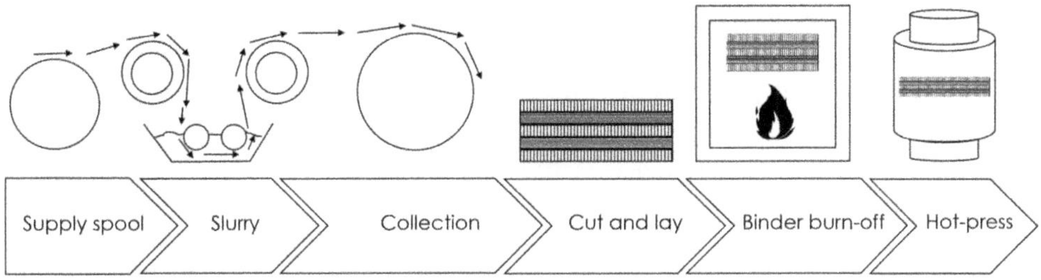

Figure 19.5. Schematic summarising SIHP of CMCs.

stable fibres being fed from a supply spool through a ceramic slurry. The slurry typically contains the matrix material in a powdered form, an organic binder, and a carrier (water/alcohol). The material quality of the final CMC is highly dependent on the slurry and as such, the slurry should be refined to match the desired outcome of the final CMC. Parameters of importance include the reinforcement diameter, particle size distribution, the matrix powder fraction, binder fraction and the type of carrier. These parameters can be used to circumvent problems arising such as voids/porosity, which can be minimised by, e.g. using matrix powder with particle sizes that are lower than the diameters of the reinforcement.

Initially, neat fibres are supplied from a spool into a slurry. Once passed through the slurry, the coated fibre is collected and dried to create a prepreg, and the ceramic powder with organic binder remains as a deposit on the fibre surfaces. The coated fibre prepregs are then cut and laid-up into stacks, after which they are initially heated to burn off the organic binder, and subsequently hot pressed, or sintered, at high temperatures and under pressure to form the final dense ceramic matrix composite. The temperature of the hot press should not exceed the melting or decomposition temperature of the reinforcement phase, whilst concurrently reaching a sufficiently elevated temperature to melt and reduce porosity in the ceramic matrix phase of the composite. The matrix phase should be soft enough to be able to flow to close pore spaces. Sintering additives can also be incorporated into the matrix if needed at the slurry stage, which will liquify during the hot pressing stage. Figure 19.5 summarises SIHP of CMCs.

19.4 Substructure/grain strengthening (MMCs)

Substructure strengthening mechanisms in metals, metal alloys and metal matrix composites (MMCs) can be most simply seen as originating from a number of interrelated factors including: dislocations, slip boundaries and grain sizes. This section will detail methods by which means strengthening of the metal matrix can be achieved at the substructural level.

19.4.1 Dislocation flow and strengthening by dislocation pile-up

Dislocations or dislocation lines, are defects in metal crystals whereby a part plane of atoms within the crystal structure is missing or there is misalignment along a plane such that atoms are anomalously arranged across the plane. The consequence of either form of dislocation line is irregular interatomic spacing that allows the dislocation to move under load, resulting in metal matrix yielding. If there is a sufficient number of dislocations, they may cross one another, which causes dislocation pinning. Pinned dislocations are unable to move as easily as free dislocations, and the consequence is that higher levels of mechanical energy are required to enable dislocation motion, resulting in metal matrix strengthening. As such, dislocation density is related to strengthening in a metal matrix and the concept can be summarised by equation (19.7), where τ is the shear flow stress, α is a constant equal to 0.5, G is the shear modulus, $\mathbf{b}$ is the Burger's vector and ρ is the dislocation density [5]. The strengthening mechanism is known is dislocation pile-up.

$$\tau = \alpha G \mathbf{b} \sqrt{\rho} \tag{19.7}$$

19.4.2 Dispersion strengthening

Metal matrices can be strengthened by the introduction of a second phase into the metal matrix by means of an alloying element. The second phase will appear as precipitated material within the metal matrix and these act as dislocation pins by physically creating a barrier to dislocation movement. This type of strengthening, known as dispersion strengthening or particulate strengthening, is most effective when the second phase particles/precipitates are dispersed homogeneously throughout the metal matrix such that the majority of dislocations can be caught and pinned during loading. The approach is different to the Orowan strengthening mechanism in MMCs as the second phase precipitates in dispersion strengthened MMCs are typically smaller than particulates involved in Orowan strengthening. Moreover, dispersion strengthened metal matrices retain their strengthening characteristics at elevated temperatures, which is not true of Orowan strengthened meta-matrices.

19.4.3 Grain boundary strengthening

A grain boundary is caused by a variation in crystallographic orientation. When the boundaries conjoin into continuous loops, they create grains or cells within the metal, which are known to affect the strength of metal. Grains/grain boundaries improve the strength of metal matrices by impeding dislocation motion. This is primarily because dislocation motion is blocked by the physical presence of the grain/grain boundary and dislocations are forced to change motive direction. Grains also create discontinuities in mobile slip planes, requiring additional levels of mechanical energy to surpass the extant impediment to slip. Plastic deformation in a metal matrix occurs when a dislocation moves from one grain to another. Since the grain boundary itself impedes dislocation motion, the higher the grain boundary density in a metal matrix, the harder it is for the dislocation to flow through each grain. Thus, the creation of smaller grains in a metal matrix will positively improve the strength of the metal matrix, as a result of the increases in boundary density. The increase of metal matrix yield strength, σ_y with grain diameter, d can be understood through the Hall–Petch relationship, equation (19.8), where σ_i is the original yield stress of the metal and k is a measure of the relative hardening contribution from grain boundaries (sometimes also referred to as the Hall–Petch slope).

$$\sigma_y = \sigma_i + \frac{k}{\sqrt{d}} \tag{19.8}$$

Grain boundaries can also be strengthened by the addition of alloying elements that locate themselves at grain boundaries. These interstitial alloying elements pin the grain boundaries, impeding their movement, and forcing the metal matrix to absorb higher levels of mechanical energy to onset grain boundary slip.

References

[1] Li T and Wang L 2017 Bending behavior of sandwich composite structures with tunable 3D-printed core materials *Compos. Struct.* **175** 46–57

[2] ASTM C393/C393M-20 2020 *Standard Test Method for Core Shear Properties of Sandwich Constructions by Beam Flexure* (West Conshohocken, PA: ASTM International)

[3] ASTM D7249/D7249M-20 2020 *Standard Test Method for Facesheet Properties of Sandwich Constructions by Long Beam Flexure* (West Conshohocken, PA: ASTM International)

[4] Wiegand N and Mader E 2017 Commingled yarn spinning for thermoplastic/glass fiber composites *Fibers* **5** 26

[5] Thompson A W 1977 Substructure strengthening mechanisms *Matell. Trans.* A **8** 833–42

IOP Publishing

Composites Engineering: An A–Z Guide

Parvez Alam

Chapter 20

'T'

20.1 Tri-axial non-crimp fabric

Tri-axial non-crimp fabrics are part of a larger family of multi-axial fabrics with non-crimp properties. The reason why these fabrics are non-crimping is because each layer consists of continuous parallel structural fibres, connected by non-structural stitching. While engineering fibres such as glass, carbon or aramid, may be used as the structural fibres, the stitching fibres are often made from polyester as it is lower in cost and has good fibre binding properties. This set-up removes the need for weaving and hence removes the crimp from the fabric. Lowering crimp results in enhanced mechanical performance as the fibres are straighter.

Individual fabric layers can be stacked to create fabrics with structural fibres oriented at different angles. Tri-axial fabrics are typically interlaced using [+45/0/−45], [+45/90/−45] or [+60/0/−60] orientations. A benefit of non-crimping methods when constructing tri-axial fabrics, is that thick fabrics can be made, and with multiple fibre orientations, reducing therefore, the need for extensive stacking to build a laminate sequence. Figure 20.1

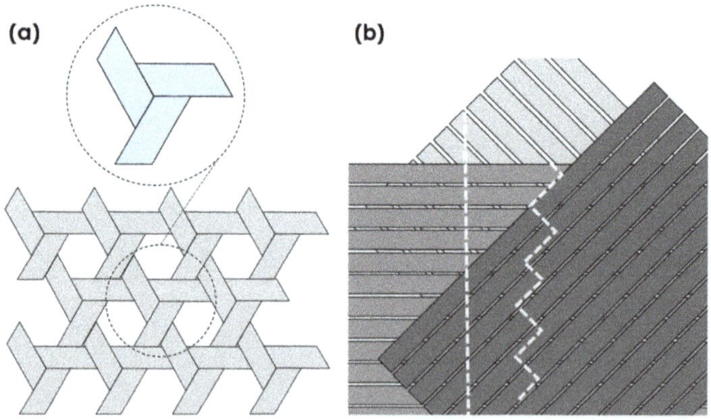

Figure 20.1. Comparison of patterns and structures of tri-axial fabrics that are (a) traditionally braided (highlighting the repeating weave pattern) and (b) non-crimping layered at [+45/0/−45] and stitched (white dashed lines).

doi:10.1088/978-0-7503-3527-0ch20

compares a tri-axial braided structure with the stitching lay-up of a tri-axial non-crimp structure. Both weave and stitch methods and simultaneous stitch methods are used to bind unidirectional structural fibres.

IOP Publishing

Composites Engineering: An A–Z Guide

Parvez Alam

Chapter 21

'U'

21.1 Unbalanced laminates

In composite laminate theory (CLT), ply stiffness can be assumed to be under a state of plane stress, equation (21.1), where Q_{ij} is termed the reduced stiffness, and σ_1 and σ_2 are normal stresses, σ_6 is shear stress, ε_1 and ε_2 are normal strains, and ε_6 is shear strain. Q_{ij} is related to stiffness, C_{ij} as $Q_{ij} = C_{ij} - \frac{C_{i3}C_{j3}}{C_{33}}(i, j = 1, 2, 6)$. σ for a 0° lamina can be calculated thence as $\sigma_i = Q_{ij}\varepsilon_j$.

$$\begin{Bmatrix} \sigma_1 \\ \sigma_2 \\ \sigma_6 \end{Bmatrix} = \begin{bmatrix} Q_{11} & Q_{12} & 0 \\ Q_{21} & Q_{22} & 0 \\ 0 & 0 & Q_{33} \end{bmatrix} \begin{Bmatrix} \varepsilon_1 \\ \varepsilon_2 \\ \varepsilon_6 \end{Bmatrix} \tag{21.1}$$

Q_{ij} relates to engineering constants as: $Q_{11} = \frac{E_{11}}{1 - \nu_{12}\nu_{21}}$; $Q_{12} = \nu_{21}Q_{11}$; $Q_{22} = \frac{E_{22}}{1 - \nu_{12}\nu_{21}}$; $Q_{16} = 0$; $Q_{26} = 0$; $Q_{66} - G_{12}$, where E is elastic modulus, ν is Poisson's ratio, and G is shear modulus.

Ply angles in laminates are often varied. Calculating the reduced stiffness so as to take into account the ply angle, $\bar{Q}_{ij}$, requires consideration of the effects of these angles on the properties, equation (21.2), where $m = \cos(\theta)$ and $n = \sin(\theta)$, θ being the ply orientation angle.

$$\begin{Bmatrix} \bar{Q}_{11} \\ \bar{Q}_{22} \\ \bar{Q}_{12} \\ \bar{Q}_{66} \\ \bar{Q}_{16} \\ \bar{Q}_{26} \end{Bmatrix} = \begin{bmatrix} m^4 & n^4 & 2m^2n^2 & 4m^2n^2 \\ n^4 & m^4 & 2m^2n^2 & 4m^2n^2 \\ m^2n^2 & m^2n^2 & m^4 + n^4 & -4m^2n^2 \\ m^2n^2 & m^2n^2 & -2m^2n^2 & (m^2 - n^2)^2 \\ mn^3 & -mn^3 & n^3m - m^3n & 2(n^3m - m^3n) \\ mn^3 & -m^3n & m^3n - n^3m & 2(m^3n - n^3m) \end{bmatrix} \begin{Bmatrix} Q_{11} \\ Q_{22} \\ Q_{12} \\ Q_{66} \end{Bmatrix} \tag{21.2}$$

$\bar{Q}_{ij}$ can then be used to determine the extensional stiffness (A_{ij}), equation (21.3), the coupling stiffness (B_{ij}), equation (21.5), and the bending stiffness (D_{ij}), equation (21.4), where t_k is the thickness of the kth lamina and $\bar{z}_k$ is the distance from the mid-plan to the centroid of the kth lamina.

$$A_{ij} = \sum_{k=1}^{n} \left(\bar{Q}_{ij}\right)_k t_k \tag{21.3}$$

$$B_{ij} = \frac{1}{2} \sum_{k=1}^{n} \left(\bar{Q}_{ij}\right)_k t_k \bar{z}_k \tag{21.4}$$

$$D_{ij} = \frac{1}{3} \sum_{k=1}^{n} \left(\bar{Q}_{ij}\right)_k \left(t_k \bar{z}_k^{-2} + \frac{t_k^3}{12}\right) \tag{21.5}$$

These in turn can then be used to calculate the forces, N, and moments, M, according to equation (21.6), which is shown in the expanded form in equation (21.7)

$$\begin{Bmatrix} N \\ M \end{Bmatrix} = \begin{bmatrix} A & B \\ B & D \end{bmatrix} \begin{Bmatrix} \varepsilon^0 \\ \kappa \end{Bmatrix} \tag{21.6}$$

$$\begin{Bmatrix} N_{xx} \\ N_{yy} \\ N_{xy} \\ M_{xx} \\ M_{yy} \\ M_{xy} \end{Bmatrix} = \begin{bmatrix} A_{11} & A_{12} & A_{16} & B_{11} & B_{12} & B_{16} \\ A_{12} & A_{22} & A_{26} & B_{12} & B_{22} & B_{26} \\ A_{16} & A_{26} & A_{66} & B_{16} & B_{26} & B_{66} \\ B_{11} & B_{12} & B_{16} & D_{11} & D_{12} & D_{16} \\ B_{12} & B_{22} & B_{26} & D_{12} & D_{22} & D_{26} \\ B_{16} & B_{26} & B_{66} & D_{16} & D_{26} & D_{66} \end{bmatrix} \begin{Bmatrix} \varepsilon_{xx}^0 \\ \varepsilon_{yy}^0 \\ \varepsilon_{xy}^0 \\ \kappa_{xx} \\ \kappa_{yy} \\ \kappa_{xy} \end{Bmatrix} \tag{21.7}$$

Here, κ_x, κ_y, and κ_{xy} are the laminate curvatures calculated as $\frac{1}{r}$, where r is the radius of curvature.

Unbalanced laminates, also referred to as asymmetric laminates or anisotropic lay-up, are laminated structures that will shear, twist and bend under the conditions of uniaxial loading. While in a balanced laminate there is geometrical symmetry, orientation symmetry and symmetry in the properties, about the centre-line, unbalanced laminates are asymmetrical in either or all of geometry, orientation and properties about the centre-line. Certain structures are best designed using unbalanced laminates. Aircraft wings and wind turbine blades [1] are good examples as the design of these using unbalanced laminates allows the wing to effectively counter coupled bend-twist deformations under aero-elastic loading. Using CLT as described above, both transverse deflection, w, and in-plane displacements at the mid-plane of the laminate u can be solved for the in-plane Cartesian coordinate systems (x, y) for cylindrical bending problems, equations (21.8) and (21.9), where N_x is specifically, the in-plane force per unit length in x and a is the half-span of the laminate.

The maximum transverse deflection, w_{max}, can be solved according to equation (21.10) where it can be noted that $w_{max} \rightarrow \infty$ as $N_x \rightarrow \infty$. Strain at the mid-plane, ε_x, can then be solved according to equation (21.11).

$$w(x) = \frac{1}{2}\frac{B_{11}N_x}{D_{11}A_{11} - B_{11}^2}\left(x^2 - a^2\right) \tag{21.8}$$

$$u(x) = \frac{D_{11}N_x}{A_{11}D_{11} - B_{11}^2} \cdot x \tag{21.9}$$

$$w_{max} = \frac{1}{2}\frac{B_{11}N_x a^2}{B_{11}^2 - A_{11}D_{11}} \tag{21.10}$$

$$\varepsilon_x = \frac{D_{11}N_x}{A_{11}D_{11} - B_{11}^2} \tag{21.11}$$

$$w(x) = \frac{q_o}{24}\left(x^4 - a^4\right) - \frac{3D_{11}A_{11} - 2B_{11}^2}{12D_{11}A_{11}}q_o a^2\left(x^2 - a^2\right) \tag{21.12}$$

$$u(x) = \frac{B_{11}}{6A_{11}}q_o x^3 - \frac{B_{11}}{A_{11}}q_o a^2 x \tag{21.13}$$

If transversely loaded, $w(x)$ and $u(x)$ can be solved for cylindrical bending according to equations (21.12) and (21.13), assuming the following pinned boundary conditions: $w(\pm a) = 0$; $M_x(\pm a) = 0$, $u(\pm a) = 0$, and where $q_o = \frac{q}{D_{11} - \frac{B_{11}^2}{A_{11}}}$ and q is the transverse load [2].

References

[1] Roczek A 2009 *Optimization of Material Layup for Wind Turbine Blade Trailing Edge Panels*

[2] Sun C T and Chin H 1988 Analysis of asymmetric composite laminates *AIAA J.* **26** 714–8

IOP Publishing

Composites Engineering: An A–Z Guide

Parvez Alam

Chapter 22

'V'

22.1 Vacuum-assisted resin transfer moulding (VARTM)

Vacuum-assisted resin transfer moulding (VARTM) is a resin infusion (RI) process similar to resin transfer moulding (RTM), except that a vacuum is used to assist resin transfer through dry reinforcing (such as a preform). While in RTM the preform is situated between two mould halves, in VARTM, the upper mould half is in fact the vacuum bag and resin flow is driven by the vacuum, figure 22.1. VARTM is a relatively low-cost processing method, which can be used to fabricate large composite parts with high volume fractions (60 %–70 %) of reinforcement. Examples of large-scale products that can be manufactured using VARTM include: aircraft landing gear doors, wind turbine blades, aircraft fuselage sections, and more. Another advantage of the process is that due to the vacuum, air is removed (debulking) from the preform and air bubble voids can be significantly reduced from the resin.

There are three main stages to VARTM. The first relates to the preparation of materials (also the pre-filling stage), the second is an infusion stage (also, filling stage), and the final is the post-infusion stage (also, post-filling stage). Each stage is summarised below:

- **Stage I—Preparation of materials:** In this stage, reinforcing material (e.g. the preform) is laid into the open mould and may be subjected to vacuum drying to eliminate moisture that may have entered the material. The purpose for drying out the preform is to ensure that entrapped moisture does not vapourise during curing, as this would lead to increased fractions of air bubble void in the final composite material. Degassing mixed resin is also important to perform before the infusion process, and this will normally be under slightly elevated temperatures and under vacuum. Elevated temperatures decrease the viscosity of resin and entrapped gasses can more easily be liberated from the resin. Debulking of the dry preform is an important consideration as it directly affects the final fibre volume fraction, the final

doi:10.1088/978-0-7503-3527-0ch22

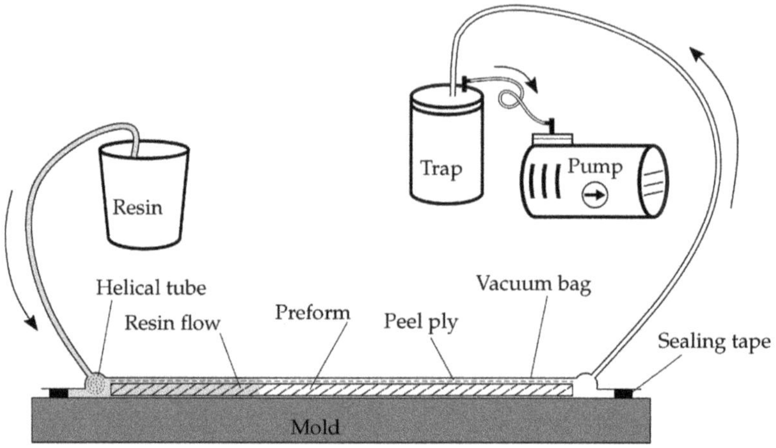

Figure 22.1. Example of a VARTM set-up. Reprinted from [1] with permission from MDPI (CC BY 4.0).

thickness of the composite, and the permeability of the preform. The leak rate is a final consideration at this stage and a typical leak rate for bag acceptance in industrial VARTM is $\leqslant 700$ Pa min^{-1} [1].

- **Stage II—Infusion:** The infusion stage is influenced by several aspects of Stage I, e.g. debulking, moisture removal, reinforcing preform material and architecture, etc. In this stage, the inlet to the resin/hardener is opened and the resin fills the preform from the resin inlet to the outlet towards the trap. The trap collects resin and preserves the longevity of the pump. The vacuum bag is overlaid and replaces the upper mould of an RTM process. The vacuum can be used during the infusion stage to minimise the thickness gradient, to vary the rate of resin flow through the preform, and to reduce the fraction of air bubble voids. Once the resin fully impregnates the preform, the infusion stage is complete.
- **Stage III—Post-infusion:** During the post-infusion stage, excess resin bleeds through the outlets, and curing takes place while remaining under a vacuum pressure [2].

22.2 Volatile organic compounds (VOCs)

Volatile organic compounds (VOCs) include a range of chemicals that can negatively contribute to human health and the environment. During the manufacture of polymer matrix composites, VOCs are often emitted during the resin curing stage, where monomeric cross-linking agents evaporate into the surrounding air from fresh resin matrix surfaces. Evaporating solvents during cleansing processes associated with composites manufacturing (e.g. tools, moulds) are also a source of VOC emission, as are wood composites containing volatile compounds that are released slowly over time (e.g. formaldehyde in plywood). The extent to which VOCs are released depends on various factors, amongst which are included, the surface area of exposure and the exposure time. Certain composites processes are naturally more VOC emitting than others. Spray lay-up for example, involves the spraying of resin over large surface areas in an atomised form, which statistically increases the opportunity for monomer evaporation and thence, VOC emission. In contrast, closed moulding operations have a very low exposed surface between the resin regions of the composite and the air, and as such there is less space available for monomer release into the surrounding air. VOC emission phenomena can loosely be said to follow Fick's second law of diffusion, equation (22.1), where $\frac{\partial C_a}{\partial t}$ is the rate of change in compound a, D_a is the diffusion coefficient of compound a, t is time, and C_a is the concentration of compound a in the surrounding air.

$$\frac{\partial C_a}{\partial t} = D_a(\nabla^2 C_a) \tag{22.1}$$

The simplest manner by which emissions, E_{mm}, can be calculated is shown in equation (22.2), where EF is the emission factor and ξ is the throughput. EF is represented by equation (22.3), where C is the chamber's air concentration, n is the ventilation air change and L is the loading factor.

$$E_{mm} = \xi \times EF \tag{22.2}$$

$$EF = C \cdot \frac{n}{L} \tag{22.3}$$

$$EF(t_i) = \frac{\left(\frac{\Delta C_i}{\Delta t_i} + NC_i\right)}{L} \tag{22.4}$$

The ASTM D 5116-06 [3] standard provides standardised guidance for the calculation of EF of materials in an environmental chamber, equation (22.4). Here, $EF(t_i)$ is the emission factor at time t_i, $\frac{\Delta C_i}{\Delta t_i}$ is the gradient of the time–concentration curve at time t_i, N is the air change rate, L is the chamber loading factor, and C_i is the chamber VOC concentration at a specified time increment [4].

22.3 Volume fraction

The volume fraction of composite components is an oft-used and important physical parameter as it governs the properties and behaviours (e.g. mechanical, electrical) of the composite. There are a variety of composite types and the method for determining or calculating volume fraction will depend to a large extent on the type of composite. In a two-component (or two-phase) composite, the volume fraction, V_f, of the reinforcing phase, $V_{f,r}$, is simply calculated according to equation (22.5), which takes into account the total volume of the composite v_c and the total volume of the reinforcing phase, v_r. The volume fraction of the matrix phase, $V_{f,m}$, is then simply expressed according to equation (22.6).

$$V_{f,r} = \frac{v_r}{v_c} \tag{22.5}$$

$$V_{f,m} = 1 - V_{f,r} \tag{22.6}$$

There are three conventional methods by which means the composite volume fraction can be determined. These include: image analysis, chemical digestion, and thermal decomposition. Image analysis methods are acceptable for continuous fibre reinforced composites where fibres are assumed to be geometrically perfect cylindrical solids that run through the entire composite, such that the area fraction of fibres can be theoretically converted to a volume fraction by multiplying out against the span of the composite. Both solvent/acid digestion methods and thermal decomposition methods involve the removal of material to enable the weighing of the reinforcing phase. The volume fraction of the reinforcing phase can then be calculated from their weight fractions, f_w, using the density, ρ, of the reinforcing and matrix phases, equations (22.7), where $_r$ and $_m$ indicate the reinforcing and matrix phases, respectively, and the volume fraction of the matrix phase can then be calculated in the same way as mentioned previously using equation (22.6).

$$V_{f,r} = \frac{f_w}{f_w + (1 - f_w)\frac{\rho_r}{\rho_m}} \tag{22.7}$$

Image analysis, chemical digestion, and thermal decomposition are discussed in more detail below.

22.3.1 Image analysis

The determination of composite volume fractions using image analysis techniques can be beneficial as it is a simple method for which both commercial and open-source tools are readily available. This method can also reduce the need for solvent digestion of matrix material, or thermal decomposition/matrix melting techniques. The technique depends not only on the resolution of the imaging tool [5], but also on the clarity and contrast between the fibre and matrix, and the accuracy of the fitting algorithm to the edges between phases. Edge tracking algorithms are often used to differentiate the fibres from the matrix, however, errors can arise when fibres are touching, or in very close proximity. If the contrast between the fibre and matrix is not sufficiently high, errors can also arise as a result of failure in the differentiation of matrix from fibre phases.

22.3.2 Solvent/acid digestion

Solvents and acids can be used to chemically digest matrix material from a composite to leave the reinforcing phase, which can be weighed and converted to a volume fraction. Using carbon fibre reinforced plastic (CFRP) as an example [6], the mass of composite material is first weighed. Then a Kjeldahl flask is used to contain a sample of composite, which is then placed in a fume extraction cupboard. Concentrated sulphuric acid is added and heated with the mouth of the Kjeldahl flask pointing away from the user. Hydrogen peroxide (50%) is added dropwise to the mixture until all colour is removed from the solution. Different types of matrix material require different quantities of solvent and the residual carbon reinforcement is washed with deionised or distilled water, dried and weighed to reveal the weight fraction of the reinforcement, which can then be converted to a volume fraction as shown in equation (22.7).

22.3.3 Thermal decomposition (burn-offs)

Heat can be used to remove matrix material, leaving the reinforcement, which similarly to chemical digestion can be collected, weighed and converted to a volume fraction using equation (22.7). Thermoset and thermoplastic FRPs are usually subjected to heats higher than the decomposition temperature of the matrix to effectively burn-off the matrix, leaving only the reinforcing phase (e.g. glass or carbon fibre). The heat is not sufficiently high to decompose the reinforcement and as such, provided there is good separation of the reinforcing phase from particulate ash, the method can be used to determine the volume fractions of the phases with a good level of accuracy.

References

[1] Beaumont W R and Zweben C H 2018 *Comprehensive Composite Materials II* (Amsterdam: Elsevier)

[2] Wu M S, Jin B C and Nutt S 2019 A recyclable epoxy for composite wind turbine blades *Adv. Manuf. Polym. Compos. Sci.* **5** 114–27

[3] ASTM D5116 06 2006 *Standard Guide for Small-Scale Environmental Chamber Determinations of Organic Emissions From Indoor Materials/Products* (West Conshohocken, PA: ASTM International)

[4] Piasecki M, Kostyrko K B and Goljan A 2021 The ability to control VOC emissions from multilayer building materials *Appl. Sci.* **11** 4806

[5] Waterbury M and Drzal L T 1989 Determination of fibre volume fractions by optical numeric volume fraction analysis *J. Reinf. Plast. Compos.* **8** 627–36

[6] Park S J and Seo M K 2011 *Composite characterization Interface Science and Technology* vol 18 (Amsterdam: Elsevier) pp 631–738

IOP Publishing

Composites Engineering: An A–Z Guide

Parvez Alam

Chapter 23

'W'

23.1 Wovens

Woven textiles and fabrics offer certain advantages over 2D laminated composites. While 2D laminated composites are known to provide optimal in-plane strength and stiffness, they exhibit out-of-plane weakness that is disadvantageous in a wide range of engineering applications, including wind and tidal turbine blades, aircraft stiffeners and springers, pressure vessels and in some areas of construction [1]. Woven textiles that use stiff and strong engineering fibres (e.g. glass or carbon) offer structural rigidity and these textiles effectively translate yarn stiffness and strength to composites. Weave structures (interlaced yarn filaments) are generally classified as 2D, 2.5D and 3D. Each will be discussed in further detail.

23.1.1 2D woven

In 2D woven technology, two orthogonal sets of yarn (warp and weft yarns) are interlaced to produce a textile with excellent in-plane properties. There are three basic 2D weaves used in a variety of composite engineering applications. These are the plain weave, the twill weave and the satin (sateen) weave, figure 23.1. The *plain weave* is the simplest type of weave (sometimes also referred to as 1 × 1 weave, Tabby weave or criss-cross weave). It is a weave where the weft yarn simply goes under and then over the warp yarn in an under-over configuration. The yarns in plain weaves are tightly interlaced and exhibit a higher degree of stability (i.e. the ability to maintain fibre orientation and weave angle). As a result of their high stability the weave is not very flexible and is difficult to mould around tight corners or complex contours without causing localised buckling and distortion. They do nevertheless mould well as flat sheets and around mild curvatures such as simple tubing. The plain weave has high crimp, and these can act as regions of high stress concentration and initiation points for failure. A traditional *twill weave* is woven so that one weft yarn crosses over two warp yarns and then under one warp yarn. It is

doi:10.1088/978-0-7503-3527-0ch23

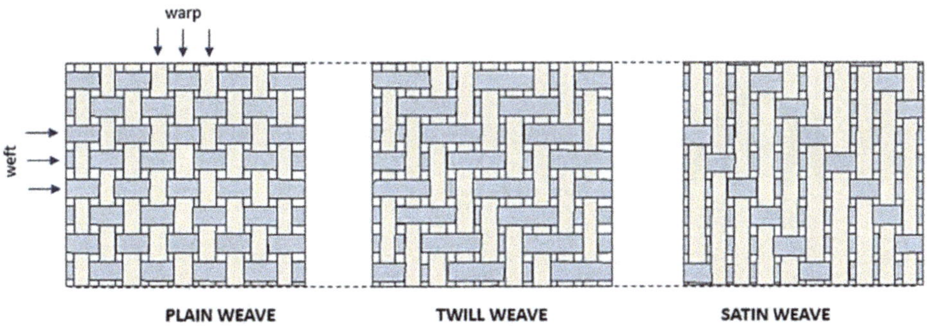

Figure 23.1. Examples of common 2D weaves identifying warp and weft yarns: plain weave, twill weave and satin weave (also called sateen weave). Reprinted from [2] with permission from MDPI (CC BY 4.0).

the most common pattern in 2D woven carbon fibre textiles however, the carbon fibre twill weave is more commonly designed in 2 × 2 and 4 × 4 weave patterns, i.e. the weft goes under two and over two warp yarns, or under four and over four warp yarns, respectively, creating a diagonal pattern. Since the interlacing distance is larger in a twill weave than a plain weave, it has lower crimp and thence, less regions of stress localisation. It should also be noted that the 4 × 4 twill weave has a lower degree of stability than the 2 × 2 twill weave, due to the difference in interlacing distance. The twill weave is more pliable than the plain weave, conforms better to contours and is therefore easier to mould into complex parts. The *satin weave*, or harness satin (HS) weave, is traditionally woven with three or more weft yarns over one warp yarn then under (3HS) or vice versa (one weft yarn over three or more warp yarns then under). Common configurations using engineering fibres such as carbon include 4HS, 5HS and 8HS, i.e. 4 under 1, 5 under 1 or 8 under 1, respectively. They have the lowest stability of the three 2D weaves described here, but they also have excellent pliability.

23.1.2 2.5D woven

In 2D woven technology, the conventional principle is to interlace two orthogonal sets of yarn (warp and weft yarns) to produce interlaced 2D textiles. A 2.5D woven technology uses a conventional 2D weaving process, however, rather than two sets of yarn (warp and weft), three sets of yarn are interlaced to manufacture pile textiles. These yarn sets are the ground warp, the pile warp and the pile weft. The ground and pile warps are warp yarns wound from separate warp beams. The ground yarns often experience a higher interlacing frequency and as such, the tension in the ground warp yarn is usually higher than that of the pile warp yarn. The looser pile warp is sometimes used to make loops, long floats, billows and other structures protuberating from the textile surface, and in such circumstances, the pile warp is longer than the ground warp. Figure 23.2 provides an example of a weaving design with tighter ground warps and looser pile warps (both face and back pile warps).

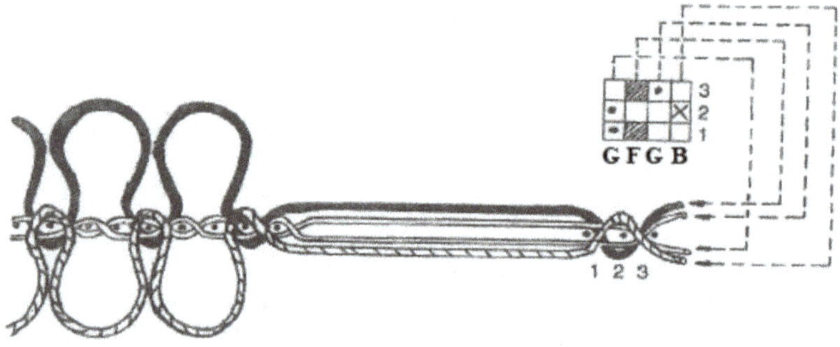

Figure 23.2. Example of a weaving design with tighter ground warps and looser pile warps (both face and back pile warps). Reprinted from [3] with permission.

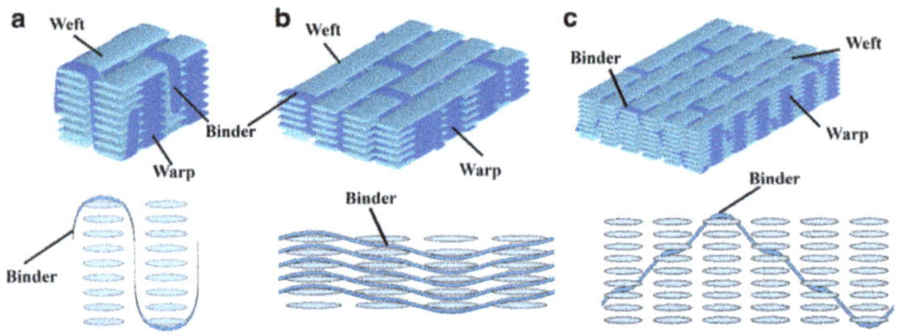

Figure 23.3. Types of 3D woven composites based on the binder path: (a) ORT, (b) LTL, (c) AI. Reprinted from [5], with permission of Springer.

23.1.3 3D woven

A 3D woven is most commonly defined as the interlacing of three orthogonal sets of yarn to produce interlaced 3D textiles with excellent in-plane *and* out-of-plane properties. The weaving shed in this instance operates on both rows and columns to produce the 3D interlaced textile. Specialty 3D weaving machines are used to enable this process, however, 2D weaves can be interlaced in the z-direction using binder warp yarn to produce certain forms of 3D textile. 2D weaving machines can also be used to create simple non-interlaced 3D textiles by weaving yarns in the warp, weft and through-thickness directions [4]. The major benefit in using a 3D woven is the enhanced out-of-plane performance of the composite, which increases its resistance to delamination, which is an essential property in many engineering applications.

3D wovens can be divided into separate groups, dependent on the depth at which the binder (z-direction) yarn penetrates through the textile. Binder yarn penetrating through the entire z-thickness is referred to as a through thickness (TT) interlock. Where binder hold only adjacent 2D layers, it is referred to as a layer-to-layer interlock (LTL). LTL can be subdivided according to the angle at which the binder yarn holds the 2D layers. If this occurs such that the binder yarn is at an orthogonal angle to the weft yarn it is termed an orthogonal interlock (ORT), whereas if the binder is at any other angle besides 90°, it is termed an angle interlock (AI). Figure 23.3(a)–(c) provides examples of the architectures of TT, LTL and AI.

The frequency at which the binder yarn is used will influence the properties of the final 3D woven including, degree of crimp, unit cell size, stiffness, and damage tolerance [5].

23.2 Wetting

The mechanical properties of any composite containing a 'wet-phase' will be directly influenced by the strength and intimacy of the bond between the components. Many factors affect the bonding between components in a composite, one of which is wetting. The spreading of liquid across a solid surface and its ability to maintain contact is wetting. Wetting is a function of the intermolecular forces of attraction at the contacting surfaces. The extent of wetting is determined by the balance between adhesive (interfacial forces) and cohesive forces of attraction within the body of fluid. Higher adhesive forces encourage spreading while higher cohesive forces make the body of liquid curl into a ball. The droplet contact angle is the angle at which three phases meet (solid, liquid and gas). The droplet contact angle is an indicator of the level of wetting, and hence the relative strengths of the adhesive and cohesive interactions. Very low contact angles indicate high wetting favourability while very high contact angles mean the liquid does not wet the solid surface very well. Figure 23.4 shows examples of non-wetting, wetting and high (near-perfect) wetting surfaces, and the contact angles that define each. Factors that influence wetting include the liquid viscosity, the surface roughness and ambient pressures.

Young's equation is often used to represent the equilibrium state of a droplet on a surface, equation (23.1), where γ is the surface energy between two phases, figure 23.5, the subscripts $_\mathrm{S}$, $_\mathrm{L}$ and $_\mathrm{G}$ refer to solid, liquid and gaseous phases, respectively, and θ_c is the contact angle.

$$\gamma_{\mathrm{SG}} = \gamma_{\mathrm{SL}} + \gamma_{\mathrm{LG}} \cos \theta_c \tag{23.1}$$

An important consideration when designing composites relates to the energy of the surface of a solid component that will be wet by a matrix material, or an adhesive. Metals and ceramics have what is said to be high energy surfaces. These materials are metallic or covalently bonded (primary bonds) and high levels of energy are required to break such bonds. Low energy surfaces are usually found on materials such as polymers, where the body of material is held together by weak secondary bonds (van der Waals). The lower the value of γ_{LG} and θ_c, the more wetting a liquid is likely to be. Good surface treatment can dramatically improve the

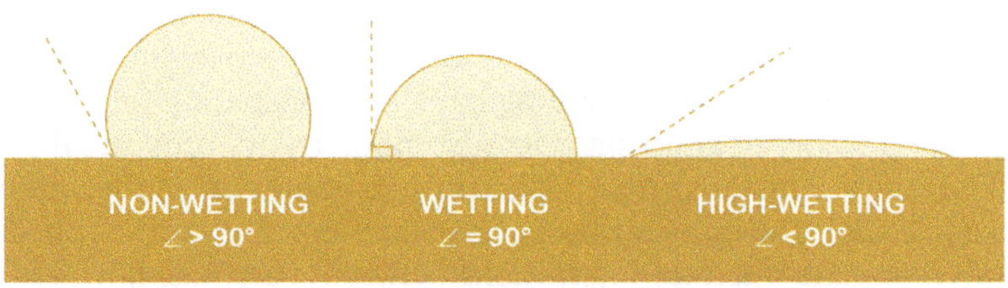

Figure 23.4. Contact angles defining different levels of wetting.

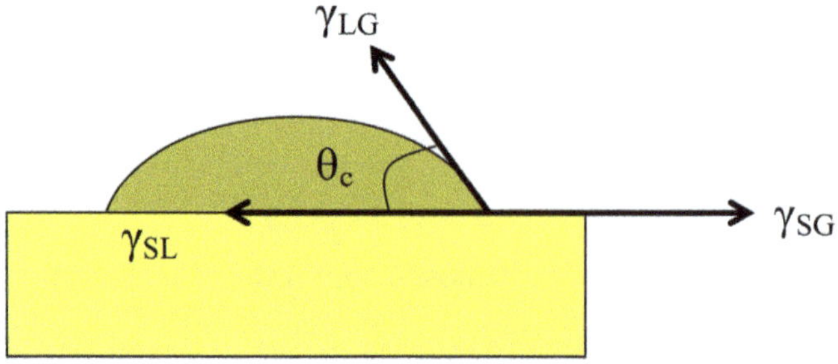

Figure 23.5. Surface energies γ between the solid (S), liquid (L) and gaseous (G) phases of a droplet on a surface.

energy level of a surface and can thence, improve the bonded contact between composite components. Steel adhesively bonded to wood or concrete for example, will need to be grit, or shot blasted, to heighten the energy of the steel surface prior to gluing. Carbon and glass fibre reinforcements, to give another example, will typically be coated in epoxy-based or silane-based sizing, respectively, to improve the wetting and thence, the strength of adhesion between the fibres and matrix in fibre reinforced plastic (FRP) composites.

With successful wetting, there will be adhesion at the interfaces of the composite components. There are a number of interdependent mechanisms that govern the final strength of bonding at these interfaces and these are discussed below:

- **Mechanical interlocking** is one such mechanism. This is essentially a topography based effect, with specific topographical profiles yielding superior interlocking characteristics. Surface preparation/treatment is nonetheless important in its own right and the effectiveness of mechanical interlocking will, to a great extent, depend upon ease of permeation into valleys, holes and general topographical irregularities. This is a function of both the liquid/matrix material and the shape of the protuberances. A mechanically roughened substrate has the benefit of increasing the surface area available for adhesion, however if there is insufficiently intimate contact, adhesion may in fact be lowered. Moreover, the shapes of certain surface irregularities may in fact increase the local stress concentrations when the composite, or the interface, is in any way strained. There are many possible surface shapes and configurations for surface irregularities. Porous materials such as wood, paper and certain large-pore ceramics have tortuous pore networks. Permeation and setting of adhesive/matrix into these networks will heighten the effects of mechanical interlocking. **Molecular entanglements** of polymer molecules at interfaces also have the effect of increasing interlocking.
- **Secondary force interactions** are perhaps the dominating interactions within adsorption theory of adhesion. The adsorption theory proposes that materials will adhere so long as there is sufficiently intimate contact between them for intermolecular forces of attraction to develop. Generally van der Waals bonds dominate interface interactions between materials, however hydrogen bonds can also develop. **Electrostatic attraction** is a secondary bonding

mechanism where oppositely charged surfaces interact. These can be dipole–dipole interactions, or, cationic–anionic couplings. **Chemical bonding** (primary bonding) may also occur across the interface, however chemisorptions as they are so called, are not as common and normally require special techniques during the process of manufacture. Bond energies for primary bonds can range from 60 to 1100 kJ mol^{-1} and 0.08 to 40 kJ mol^{-1} for secondary bonds.

- **Polymer diffusion** and migration between contacting interfaces (diffusion theory) is a means by which polymers adhere to each other (autohesion). This theory deems essential, high molecular chain mobility at interfaces for diffusion to take place alongside similar levels of solubility. Generally, long chain segments maintain position in the host polymer and chain ends will diffuse into adjacent polymeric material. The consequence is that the short segment of chain effectively anchors itself into adjacent polymer and if this process repeats sufficiently, polymer–polymer adhesion occurs. **Molecular entanglements** may form when chain segments are sufficiently mobile.
- **Interphase regions** are regions of contacting materials that have combined local composite properties of both materials. Example interphase regions could be *adhesive–wood* in jointed wood composites, *binder–paper* in coated paper composites and *polymer–polymer* in joined polymers (joined through polymer chain diffusion). The extent of influence such interphase regions will have on the composite properties will depend on a number of factors including but not limited to: depth of penetration relative to depth of composite, properties of penetrated /adsorbed material relative to substrate, intimacy of bonding between materials within the interphase, and the topographical profile.

23.3 Whiskers

Whiskers are high-cost, single-crystal, defect-free filaments of material used to reinforce polymer matrix, ceramic matrix, and metal matrix composites. They can be formed using small-scale crystallisation techniques, and the alignment of atoms within whiskers is almost perfect. They are used as discontinuous fibre composite reinforcements and since they are monocrystalline and free of defects, they have exceptional mechanical properties. Common whiskers used as reinforcing in composites include: aluminium oxide (alumina), silicon carbide, iron, hydroxyapatite, graphite, cellulose and silicon. Whiskers are essentially reinforcing fibres with extremely low aspect ratios. They are known to contribute to the final properties of the composite by increasing strength at elevated temperatures, improving the resistance to creep, thermal shock and stress rupture. A disadvantage of whiskers though, is that since they are very short fibres with very low aspect ratios, they are neither effective at transferring stress as continuous fibre reinforcements, nor are they easily oriented to a single direction, unlike continuous fibre reinforcements.

Depending on the material, the strength of whiskers can be between ca. 3 and 100 fold greater than the strength of continuous filaments made from the same material.

Aluminium oxide for example ordinarily has a tensile strength in the region of 105 MPa. In its monocrystalline (whisker) form, it is closer to 19 GPa, while the difference between continuous filament and whisker forms of silicon carbide is ca. 3.4 GPa and 11 GPa, respectively. Though they have a very high strength, they are also extremely brittle and tend to break into shorter aspect ratio fibres during processing, further lowering their stress transfer capacity when embedded within a matrix. Nevertheless, they have excellent wear resistance and a 20% fraction of silicon carbide whisker embedded within an aluminium 2024 alloy has been shown to be 6-times more wear resistant than the neat Al-2024 alloy [6].

There are disadvantages in using whiskers. One is that they are high-cost reinforcements and another is that they can sometimes be difficult to disperse homogenously within a matrix. By example, the cost of silicon carbide whiskers is typically 30-times the cost of silicon carbide particulates, both of which can be effective in improving wear/resistance in metal matrix composites. Another disadvantage is that, due to their small sizes, they can partly orient during certain processing techniques such as extrusion or rolling, where whiskers orient towards the direction of mass flow. This can cause unintended anisotropy in the final composite material, which is a desirable property in some but not all instances. A final disadvantage that shall be mentioned here relates to their shape. Since they have an aspect ratio, albeit low, they retain higher aspect ratios than particulates, which have aspect ratios closer to unity. As such, when the volume fractions of whiskers are high, they cannot pack as tightly and effectively as particulate reinforcements, which limits the versatility of their use.

23.4 Wood-based composites

23.4.1 Chipboard

Chipboard, also known as particle board or low density fibreboard (LDF), is an engineered wood product made from disposable wood chips or fibres. The chips are bound together by epoxy resin and the wood/epoxy composite is pressed under intense heat to form flat sheets. Often a three layer process is used to make chipboard. The central layer is formed of large wood chips, the outer layers are small wood chips or fibres, and between these layers are mid-sized wood chips. Nevertheless, the chipboard microstructure can be varied as needed, figure 23.6, and the choice of structure will depend on cost and end-use [7].

Figure 23.6. A variety of chipboard structures: chip-sawdust with outer layer from microchips (left), chip-sawdust (middle), and chipboards (right). Reprinted from [7] with permission from MDPI (CC BY 4.0).

23.4.2 Cross-laminated timber (CLT/X-lam)

Cross-laminated timber, also referred to as CLT or X-lam, is a composite wood panel manufactured by gluing together layers of sawn, kiln-dried lumber. The lumber pieces can vary in size but as can be seen from figure 23.7, long thin pieces of lumber can be connected together to create large structural panels. Each lumber layer is arranged orthogonally to the layer beneath and above it, and this gives the panel very good in-plane properties, and reduces dimensional changes due to shrinkage. By using cut lumber, structural anomalies such as knots and fissures are spread out and made discontinuous, thus decreasing their effectiveness in crack propagation.

23.4.3 Glue-laminated timber (Glulam)

Glued-laminated beams, or glulam, is similar to CLT in that cut dried lumber is laminated together and glued. The main difference is that glulam is not cross-laminated but rather, the lumber is stacked in parallel, figure 23.8. This optimises the strength and stiffness properties of glulam along its axis and the composite is very useful where resistance to axial bending is desired. By using cut lumber, structural anomalies such as knots and fissures are spread out and made discontinuous, thus decreasing their effectiveness in crack propagation.

23.4.4 Laminated veneer lumber (LVL)

Laminated veneer lumber, or LVL, is a structural composite lumber where thin wood veneers are laminated with the fibre axis orientation being the same in each veneer lamina. The veneers are oriented in the same direction and are adhesively bonded and hot-pressed. Major lumber defects such as knots and fissures are segmented through the process of veneering, and these are then distributed more homogenously across the LVL product after bonding. LVL is the most widely used structural composite lumber and has exceptional properties of strength, stiffness and dimensional stability when compared to the neat lumber from which it is made.

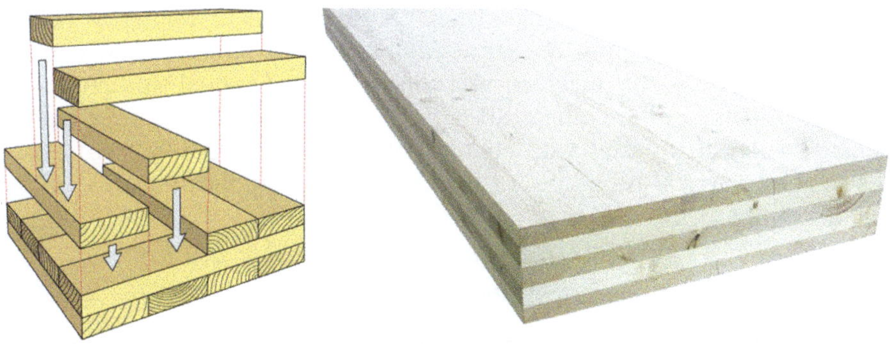

Figure 23.7. Structure and lay-up of CLT. Reprinted from [8] with permission from MDPI (CC BY 4.0).

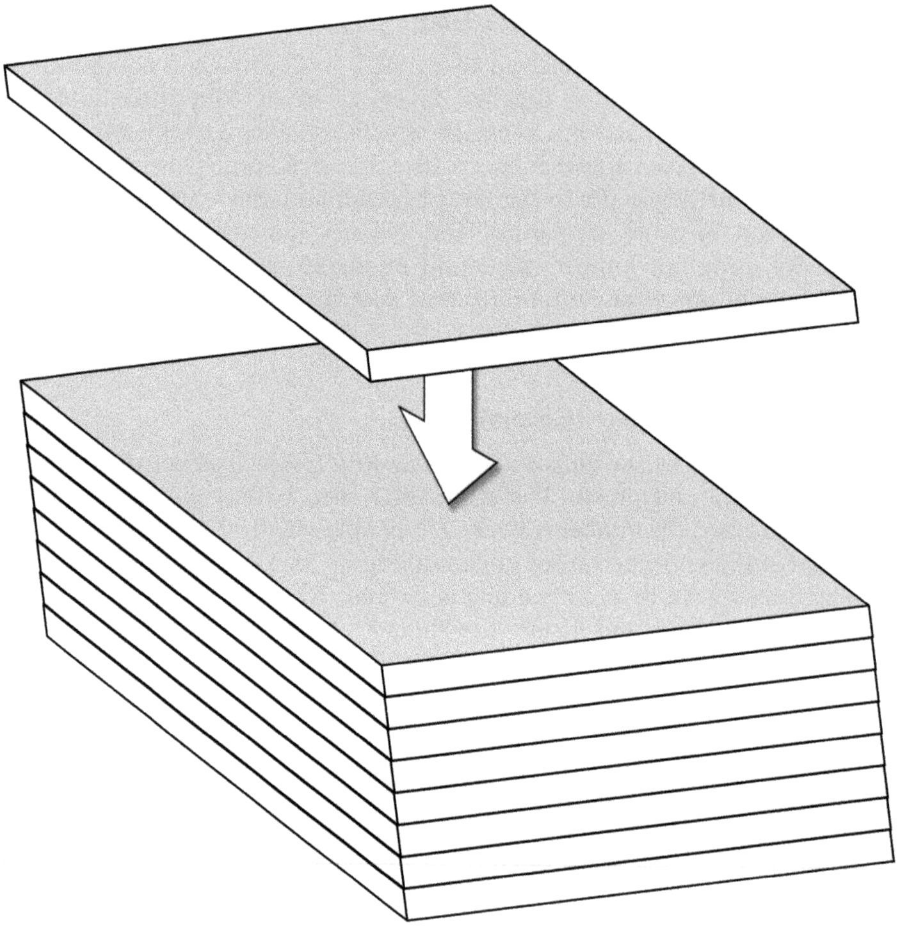

Figure 23.8. Structure and lay-up of glulam.

23.4.5 Oriented strand board (OSB)

Oriented strand board, or OSB, is made of wood chips compressed together with resin under heat to form strong flat structural panel. It is dissimilar to chipboard since in OSB, the wood chips are strategically oriented into a strand mat before pressing to maximise the mechanical properties of the final composite panel, figure 23.9. A benefit of creating OSB is that waste wood chips can be made into continuous wood composites with controllable mechanical properties. Depending on the final mechanical properties of the OSB, its end applications may either be non-load bearing or load bearing.

23.4.6 Parallel strand lumber (PSL/Parallam)

Parallel strand lumber (also PSL or parallam) is a structural composite lumber where long strands of wood (long wood chips) are glued together by epoxy. The strands are

Figure 23.9. Strand mat before pressing (left) and detail showing the surface of an OSB board (right). Reprinted from [9] with permission from MDPI (CC BY 4.0).

pressed and heated to create a final solid beam that can be used in a variety of construction applications. Each strand is laid with (as close to) a 0° fibre orientation as possible. The benefit of the process is that it enables the use of long wood chips that would otherwise be put to waste. The final PSL beam has high strength and stiffness properties along the axis of the beam (figure 23.10).

23.4.7 Plywood

Plywood is a cross-laminated engineered composite wood panel, where unlike CLT the individual lamina are veneers (not lumber) and unlike LVL the veneers are cross-laminated (i.e. with an orthogonal [0/90°] lay-up).

23.4.8 Wood–plastic composite (WPC)

Wood–plastic composites, or WPCs, are produced by mixing wood flour with thermoplastic resins. Typical thermoplastic resins used include polypropylene, polyethylene, polyvinyl chloride, or polylactic acid. A benefit of WPCs is that the filler material (the wood flour) can be easily used from waste wood chippings. WPCs have respectable properties of specific strength and stiffness when considering the

Figure 23.10. Structure of PSL. Wikipedia: Simon Mer (CC BY SA 4.0).

sizes of the reinforcement used. They are additionally, highly wear resistant. The applications of WPCs are somewhat limited by their properties and they tend to be used primarily in exterior decking material [10].

References

[1] El-Dessouky H and Saleh M N 2018 3D Woven composites: from weaving to manufacturing *Recent Developments in the Field of Carbon Fibers* ed R Khanna and R Cayumil (Rijeka: InTech)

[2] Singh C, Wong C S and Wang X 2015 Medical textiles as vascular implants and their success to mimic natural arteries *J. Funct. Biomater.* **6** 500–25

[3] Oner E, Durur G and Sirin B 2017 *IOP Conf. Ser.: Mater. Sci. Eng.* **254** 162009

[4] Soden J A and Hill J 1998 Conventional weaving of shaped preforms for engineering composites *Composites* A **29** 757–62

[5] Saleh M N and Soutis C 2017 Recent advancements in mechanical characterisation of 3D woven composites *Mech. Adv. Mater. Mod. Process.* **3** 12

[6] U S Congress, Office of Technology Assessment 1988 *Advanced Materials by Design, OTAE-351* (Washington, DC: U.S. Government Printing Office)

[7] Mirski R, Derkowski A, Dziurka D, Dukarska D and Czarnecki R 2019 Effects of a chipboard structure on its physical and mechanical properties *Materials* **12** 3777

[8] Di Bella A and Mitrovic M 2020 Acoustic characteristics of cross-laminated timber systems *Sustainability* **12** 5612

[9] Zanuttini R, Bonzano E, Negro F, Oreglia G L and Cremonini C 2020 Preliminary assessment of sweet chestnut and mixed sweet chestnut-poplar OSB *Forests* **11** 496

[10] Murayama K, Ueno T, Kobori H, Kojima Y, Suzuki S, Aoki K, Ito H, Ogoe S and Okamoto M 2019 Mechanical properties of wood/plastic composites formed using wood flour produced by wet ball-milling under various milling times and drying methods *J. Wood Sci.* **65** 5

IOP Publishing

Composites Engineering: An A–Z Guide

Parvez Alam

Chapter 24

'X'

24.1 X-ray computed tomography (x-ray CT/XCT)

X-ray computed tomography (x-ray CT, or, XCT) has become rapidly more popular amongst composites engineers as it is a non-destructive technique that enables the internal, 3-dimensional, (sometimes) micro-level detection of geometry and space in composites. The internal architectures of composites are not ordinarily observed and as such the technique opens several new opportunities to develop understanding of composite materials in processing, as manufactured, and in post-testing states. *In situ* x-ray CT is an important tool for the understanding of a wide range of time-dependent phenomena in composites, from in-service degradation to in-autoclave and out-of-autoclave processes [2].

XCT works on the basis of x-ray signal attenuation. X-rays are released to the composite from an x-ray source, the x-rays pass through the composite and detectors measure the extent to which the x-ray signals have been attenuated by the composite. Figure 24.1 (from [1]) shows an example of XCT set-ups for medium (a) and large (b) sized samples. Attenuation is a result of signal scattering and absorption, and the fundamental equation of a mono-energetic beam through a material (assuming homogeneity) is expressed through Beer's law, equation (24.1), where I_0 is the initial x-ray intensity, I is the final x-ray intensity, μ is the material linear attenuation coefficient and x is the length of the x-ray path.

$$I = I_0 \exp\left[-\mu x\right] \tag{24.1}$$

In a composite, there are multiple materials and Beer's law is then expressed by equation (24.2), where the subscript i refer to the individual material components of the composite.

$$I = I_0 \exp\left[\sum_i \left(-\mu_i x_i\right)\right] \tag{24.2}$$

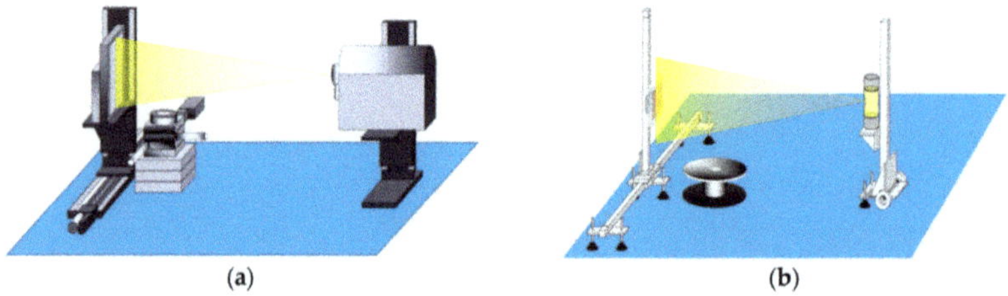

Figure 24.1. Examples of x-ray tomography set-ups for medium sized (a) and large sized (b) samples. Reprinted from [1] with permission from MDPI (CC BY 4.0).

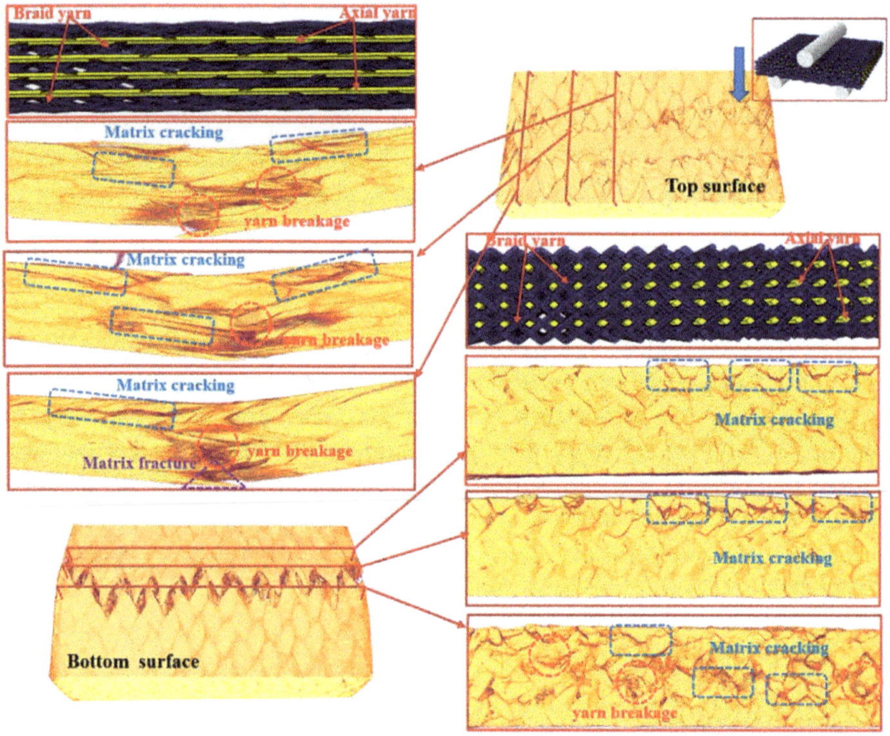

Figure 24.2. Examples of micro-scale damage in a hybrid 3D braided polymer matrix composite as detected through x-ray tomography. Reprinted from [4] with permission from MDPI (CC BY 4.0).

If a polychromatic x-ray source is used, the x-ray energy (of incident photons), E, is incorporated into equation (24.2) to give equation (24.3). There are nevertheless perceived problems with this since it assumes μ is a linear attenuation coefficient, rather than one that is absolute. Nevertheless, this is still common in the XCT of composites and μ for a specific point in a material can be expressed by equation (24.4), where (x, y, z) are Cartesian descriptors for the position of μ in Euclidean space, K is a constant, ρ is the density of the material at position (x, y, z), and Z is the atomic number of the material [3].

$$I = I_0(E)\exp\left[\sum_i(-\mu_i(E)x_i)\right]dE \tag{24.3}$$

$$\mu(x, y, z) = K\rho\frac{Z^4}{E^3} \tag{24.4}$$

While there are several advantages to XCT, it should be noted that the resolution of the computational reconstruction (typically voxel based) is limited by the size of the object. A 1000–2000 resolution of the cross-section is common, but if the cross-section is large, finer detail at the lower length scales will not be observed or reconstructed. This issue with resolution has an additional effect of blurring boundaries and large volumes of data are produced from XCT scans. Nevertheless, with careful planning and judicious decisions on the scan set-up and procedure, excellent 3D images can be reconstructed, such as the example in figure 24.2 from [4], which shows different types of damage (matrix failure and yarn breakage) in a 3D braided polymer matrix composite subjected to bending loads.

References

[1] Albertin F, Bettuzzi M, Brancaccio R, Morigi M P and Casali F 2019 X-ray computed tomography *in situ*: an opportunity for museums and restoration laboratories *Heritage* **2** 2028–38

[2] Naresh K, Khan K A, Umer R and Cantwell W J 2020 The use of x-ray computed tomography for design and process modeling of aerospace composites: a review *Mater. Des.* **190** 108553

[3] Garcea S C, Wang Y and Withers P J 2018 X-ray computed tomography of polymer composites *Compos. Sci. Technol.* **156** 305–19

[4] Wu L, Sun X, Xiang C, Wang W, Zhang F, Jiang Q, Tang Y and Lin J-H 2020 Short beam shear behavior and failure characterization of hybrid 3D braided composites structure with x-ray micro-computed tomography *Polymers* **12** 1931

IOP Publishing

Composites Engineering: An A–Z Guide

Parvez Alam

Chapter 25

'Y'

25.1 Yarn

A yarn is an assemblage of fibres arranged initially in parallel, after which a twist is applied to create a continuous length of fibrous material with unique mechanical properties. Yarns are important in textile composites, where they may be braided, woven, knitted or kept as non-wovens. Yarns are typically constructed to have either an S-twist or a Z-twist, which are essentially twists of the parallel fibres in anti-clockwise or clockwise directions, respectively. The benefits of twisting include: holding together the individual fibres in the yarn, improvement in the mechanical properties of the fibrous structure, and improvements in abrasion and wear resistance.

Lower numbers of twists in a yarn make the material loose and difficult to work with, and breakage or axial splitting in such cases is more likely. Increasing the twist in a yarn makes it stiffer and keeps the fibres together more effectively, but for each twist, there is an increased stress upon individual fibres within the yarn and this can affect the mechanical performance of textile composites that are made from these yarns. The number of twists in a yarn (degree of twist), DoT, can be calculated using equation (25.1), where n is the number of turns and l is the length over which n is counted. The output is typically referred to in terms of TPI (turns per inch), TPM (turns per metre) or TPC (turns per centimetre).

$$\mathrm{DoT} = \frac{n}{l} \tag{25.1}$$

The angle of twist, θ_{twist}, is also a mechanically important parameter, since this angle influences the mechanical behaviour of the yarn. A higher diameter yarn will naturally have a higher twist angle, and a higher twist angle reduces the stiffness of the yarn in the following two ways: (1) the optimal orientation angle for a fibre is 0° and any higher than this is a reduction of stiffness E_{reduced} recognising that $E_{\mathrm{reduced}} = E \cdot \cos\theta_{\mathrm{twist}}$, and (2) since larger diameter yarns will naturally cause the

doi:10.1088/978-0-7503-3527-0ch25

angle of twist to increase, there will be a greater fraction of the length of each fibre that will be oriented at angles >0°. θ_{twist} is calculated according to equation (25.2), where, ϕ_{yarn} is the diameter of the yarn.

$$\tan \theta_{\text{twist}} = \frac{\pi \phi_{\text{yarn}}}{l} \tag{25.2}$$

To account for both the degree of twist and the yarn diameter, a product can be formed between the TPM and the square root of the yarn *tex*, where *tex* is a measure of linear density (e.g. grams per 1000m). This product is termed the twist factor or twist multiplier, K_t and is shown in equation (25.3). K_t is a useful parameter as it can be used to determine the TPM of a yarn during spinning. It can also be used in other ways, such as to describe the different characteristics (e.g. softness, or hardness) of yarns spun in the same material with different K_t.

$$K_t = \text{TPM} \cdot \sqrt{tex} \tag{25.3}$$

Chapter 26

'Z'

26.1 Z-pinning

Z-pinning is a physical method aimed at reducing delamination in composites by reinforcing the laminate in its z-direction (through thickness direction). Z-pins are either metallic or pre-cured composite pins that are wedged between continuous fibres in the z-direction of the laminate. They are typically inclined at an offset angle to the z-direction in the region of 5°–15° and up to 20°. The method is aimed primarily at the z-reinforcement of prepreg laminates and pins are typically inserted into the uncured prepreg laminates using an ultrasonic hammer. The hammer vibrates at a high frequency, creating heat at the interface between the pin and the laminate, consequently softening the material such that the pin can sink into the prepreg with relative ease. Protruding z-pin material can be removed and the z-pinned prepreg can be cured. Pins should be judiciously placed and care should be taken to decide on an appropriate pin diameter as both placement and size can, if excessive, be of detriment to the mechanical properties of the z-pinned laminate. The pins themselves couple frictional and adhesive forces at the interface to constrain delamination during loading [1].

Z-pins in a laminate disrupt the local continuity and orientation of fibres, figure 26.1. While z-pinning improves delamination tolerance, it does introduce defects that can affect the in-plane strength, stiffness and toughness of the laminate. Local waviness is increased in the vicinity of each z-pin and on average, for each doubling of the fraction of z-pins in a structure, there is an approximate doubling of the fraction of waviness. Waviness is known to negatively affect the in-plane mechanical properties of laminates and as such, care should be taken to account for mechanical losses when introducing z-pinning into the design process.

Mechanical losses can also arise through the presence of interconnected resin rich regions (cf. Figure 26.1). Z-pinning creates these regions by separating fibres and hence leaving void space into which resin can enter. If too closely spaced, resin rich regions can interconnect in the fibre axis, figure 26.2(a) and this problem can be

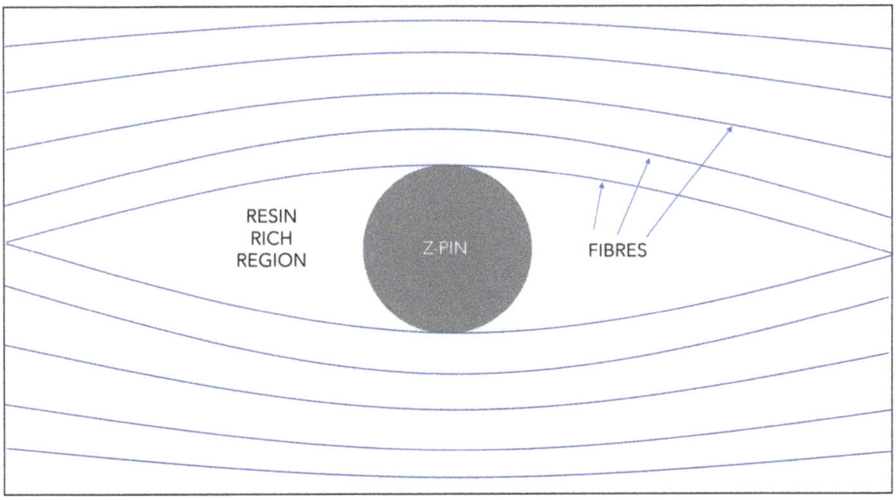

Figure 26.1. Local fibre reorientation as a consequence of z-pinning.

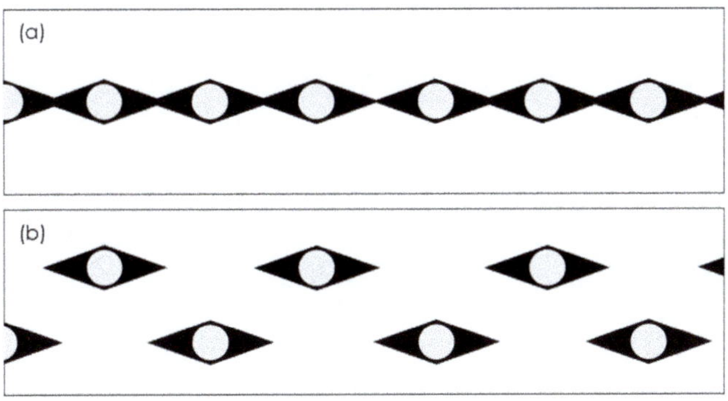

Figure 26.2. Resin rich regions (a) interconnecting by being too closely spaced along the same axis and (b) remaining separate by staggering the pins (pins shown in grey and resin rich regions shown in black).

avoided more effectively by staggering locations of the pins prior to their embedding into the procure, figure 26.2(b) [2].

While sensible fractions of z-pinning (2 %–4 % by volume) improves properties such as the interlaminar fracture toughness and impact damage tolerance, it concurrently lessens the in-plane mechanical properties of laminates such as strength, stiffness and fatigue life.

References

[1] Mouritz A P 2007 Review of z-pinned composite laminates *Composites* A **38** 2383–97

[2] Chang P, Mouritz A P and Cox B N 2007 Flexural properties of z-pinned laminates *Composites* A **38** 244–51

www.ingramcontent.com/pod-product-compliance
Ingram Content Group UK Ltd.
Pitfield, Milton Keynes, MK11 3LW, UK
UKHW051853100926
8556IPUK00005B/612